Principles and Applications of Fourier Optics

Principles and Applications of Fourier Optics

Robert K Tyson

University of North Carolina at Charlotte, USA

IOP Publishing, Bristol, UK

ISBN 978-0-750-31056-7 (ebook)
ISBN 978-0-750-31057-4 (print)
ISBN 978-0-750-31124-3 (mobi)

DOI 10.1088/978-0-750-31056-7

Version: 20140801

IOP Expanding Physics
ISSN 2053-2563 (online)
ISSN 2054-7315 (print)

British Library Cataloguing-in-Publication Data: A catalogue record for this book is available from the British Library.

Published by IOP Publishing, wholly owned by The Institute of Physics, London

IOP Publishing, Temple Circus, Temple Way, Bristol, BS1 6HG, UK

US Office: IOP Publishing, Inc., 190 North Independence Mall West, Suite 601, Philadelphia, PA 19106, USA

Contents

Preface

'All of optics is Fourier optics!' While this statement may not be literally true, when there is one basic mathematical tool to explain light propagation and image formation, with both coherent and incoherent light, as well as thousands of practical everyday applications of the fundamentals, Fourier optics is worth studying. This book contains five chapters with a summary of the principles of Fourier optics that have been developed over the past hundred years and two chapters with summaries of many applications over the past fifty years, especially since the invention of the laser. It is not necessary to take a new tack in explaining the principles. I have drawn liberally from the descriptions and notation of the authors of the classic texts on the topic, namely Professor Joseph V Goodman of Stanford University and Professor Jack D Gaskill of the University of Arizona. For applications that have evolved recently and are being reported even as this book goes to press, I have attempted to maintain the consistent notation and definitions of the early chapters, but deferred to the notation of the recent researchers where consulting the original papers with different notation would be confusing. This book is intended as a series of guideposts to improve our world through the applications of Fourier optics.

Robert K Tyson, PhD
Charlotte, NC, USA

Author biography

Robert K Tyson

Robert K Tyson is an associate professor emeritus at the University of North Carolina (UNC) at Charlotte. He has a bachelor's degree in physics from the Pennsylvania State University and MS and PhD degrees in physics from West Virginia University. Following his formal education, he worked in the aerospace industry for 20 years designing optical systems and supporting technology for strategic-defense high-energy laser weapon systems. He is the author of seven books on adaptive optics at various reading and interest levels. Dr Tyson joined the faculty of UNC Charlotte in 1999.

IOP Publishing

Principles and Applications of Fourier Optics

Robert K Tyson

Chapter 1

Theory of optical propagation and diffraction

1.1 The Helmholtz equations

1.1.1 The wave equation in the time domain

We start with Maxwell's equations which describe the interaction between an electric field vector $\boldsymbol{E}(E_x, E_y, E_z)$ and a magnetic field vector $\boldsymbol{H}(H_x, H_y, H_z)$.

$$\begin{aligned} \nabla \times \boldsymbol{E} &= -\mu \frac{\partial \boldsymbol{H}}{\partial t} \\ \nabla \times \boldsymbol{H} &= \varepsilon \frac{\partial \boldsymbol{E}}{\partial t} \\ \nabla \bullet \varepsilon \boldsymbol{E} &= 0 \\ \nabla \bullet \mu \boldsymbol{H} &= 0 \end{aligned} \tag{1.1}$$

This assumes that there is no free charge; μ is the permeability and ε the permittivity of the propagation medium. Both electric and magnetic field vectors are functions of both position and time, represented by their Cartesian components and the time derivative.

The equations can be combined by taking the cross-product of the first two time-derivative equations, which results in the wave equations.

$$\begin{aligned} \nabla^2 \boldsymbol{E} - \frac{n^2}{c^2}\frac{\partial^2 \boldsymbol{E}}{\partial t^2} &= 0 \\ \nabla^2 \boldsymbol{H} - \frac{n^2}{c^2} H &= 0 \end{aligned} \tag{1.2}$$

We assume that the medium is linear, isotropic, non-dispersive and homogeneous. That is, ε and μ are constant. We define ε_0 and μ_0 as the constants in a vacuum. The refractive index is $n = (\frac{\varepsilon}{\varepsilon_0})^{\frac{1}{2}}$ and the velocity of propagation in a vacuum is $c = \frac{1}{\sqrt{\varepsilon_0 \mu_0}}$. The vector equations are obeyed by both the electric and magnetic fields and there is a scalar equation that is obeyed by each component of the fields. The expression for the x-component of the electric field equation becomes

$$\nabla^2 E_x - \frac{n^2}{c^2}\frac{\partial^2 E_x}{\partial t^2} = 0. \tag{1.3}$$

Now we can generalize to any scalar field component $u(x, y, x, t)$

$$\nabla^2 u(x, y, x, t) - \frac{n^2}{c^2}\frac{\partial^2 u(x, y, x, t)}{\partial t^2} = 0. \tag{1.4}$$

Reducing the vector equations to scalar equations has consequences. If a medium is inhomogeneous where ε depends on position, a coupling occurs between the components of the fields and all components will not necessarily satisfy the same wave equation. Furthermore, at an interface between two media, the boundary conditions must be satisfied over a small region where an error occurs between the vector results and the scalar approximation. This is the crux of the study of Fourier optics. By making more simplifying assumptions about the propagation conditions, which result in simpler and more solvable equations, errors in the details appear. By limiting the errors, we can use simpler expressions and many useful mathematical tools to provide practical results.

Instead of using the Cartesian coordinates x, y, z, we can combine the spatial coordinates into one three-dimensional coordinate, s. Let us assume that we have a monochromatic wave that must satisfy the wave equation. The wave number k is given by $k = \frac{2\pi}{\lambda}$ where λ is the wavelength. And, where the velocity in the homogeneous dielectric medium is ν, the wavelength is defined as $\lambda = \frac{c}{n\nu}$. One solution that represents a traveling wave is

$$u(s, t) = \mathrm{Re}[U(s)\exp(-\mathrm{j}2\pi\nu t)] = A(s)\cos[2\pi\nu t - \phi(s)]. \tag{1.5}$$

$A(s)$ is the wave amplitude and $\phi(s)$ is the wavefront phase. Knowing that the Laplacian is

$$\nabla^2 u = \sum_{i=1}^{3}\frac{\partial^2 u}{\partial s_i^2}. \tag{1.6}$$

Equation (1.5) satisfies the time independent wave equation

$$(\nabla^2 + k^2) = 0. \tag{1.7}$$

There are many solutions to the wave equation. Waves can be traveling spherically from a source point. They can be plane waves where the plane is normal to the direction of propagation. Depending upon the boundary conditions and source of the radiation, they can take on many different forms. In 1678 Christian Huygens stated that any wavefront of a field that satisfies the wave equation can be considered to be made up of an infinite number of spherical source points that form a new wavefront as it propagates. See figure 1.1.

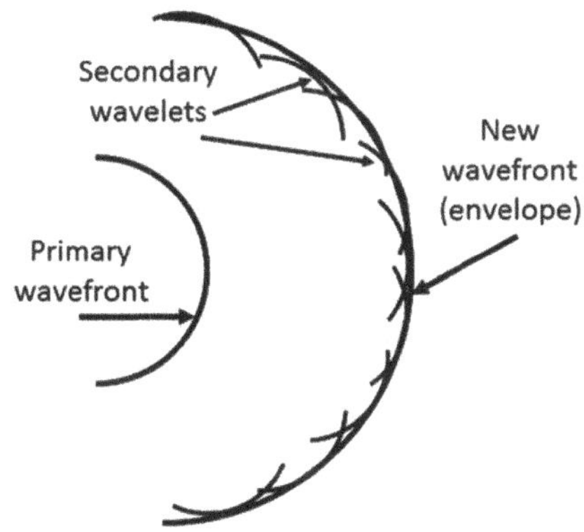

Figure 1.1. Huygens' spherical wavelets form a new wavefront.

1.2 The integral theorem of Helmholtz and Kirchhoff

We quantify this by solving the spherical wave problem using Green's theorem. Most calculus texts cover the proof and general application of the theorem, but because I absolutely abhor rigorous mathematical proofs, I will avoid repeating it here. It is stated simply as an equality between the energy leaving a volume being equal to the energy leaving a surface enclosing the volume in a direction normal to that surface:

$$\iiint_V (U\nabla^2 G - G\nabla^2 U)\mathrm{d}V = \iint_S \left(U\frac{\partial G}{\partial n} - G\frac{\partial U}{\partial n}\right)\mathrm{d}S. \tag{1.8}$$

The partial derivatives are in the direction of an outward normal to the surface. The point of observation is P_0. The surface S surrounds the point P_0. Kirchhoff chose the so-called free space Green's function to be a spherical wave expanding about the point P_0.

The Green's function at any point P_1 at a distance r_{01} from P_0 is

$$G(P_1) = \frac{\exp(jkr_{01})}{r_{01}} \tag{1.9}$$

and

$$\frac{\partial G(P_1)}{\partial n} = \cos(\boldsymbol{n}, \boldsymbol{r}_{01})\left(jk - \frac{1}{r_{01}}\right)\frac{\exp(jkr_{01})}{r_{01}}. \tag{1.10}$$

The cosine term is the cosine of the angle between the outgoing normal vector $\boldsymbol{n}$ and the vector $\boldsymbol{r}_{01}$.

After a page or so of mathematics found elsewhere (Goodman 2005) the result is an integral expression for the field at P_0 as a function of the field at P_1. The result, known as the integral theorem of Helmholtz and Kirchhoff (Kirchhoff 1883) is

$$U(P_0) = \frac{1}{4\pi}\iint_S \left\{\frac{\partial U}{\partial n}\left[\frac{\exp(jkr_{01})}{r_{01}}\right] - U\frac{\partial}{\partial n}\left[\frac{\exp(jkr_{01})}{r_{01}}\right]\right\}\mathrm{d}S. \tag{1.11}$$

The choice of the surface or surfaces of integration is critical to the validity of the result. In many cases involving practical diffraction problems, one of the surfaces will be a sphere of some radius R'. For the integrals to be performed, we want to guarantee that we are dealing with only an outgoing wave through the surface. We have to ensure that the field U vanishes at least as fast as the diverging spherical wave. To do so, we invoke the Sommerfeld radiation condition

$$\lim_{R'\to\infty} R'\left(\frac{\partial U}{\partial n} - \mathrm{j}kU\right) = 0. \tag{1.12}$$

1.2.1 The Kirchhoff boundary conditions

Let us now assume that we have a propagating electromagnetic field impinging on an opaque screen where diffraction occurs due to some opening in the screen. We want to calculate the field at points behind the screen. To use the formulation so far described, we have to assert two conditions called the Kirchhoff boundary conditions. The first is that the surface of integration after the screen is entirely arbitrary and the field U and its derivative $\frac{\partial U}{\partial n}$ are exactly the same as they would be in the absence of the screen. The second condition is that the screen is really opaque, that is, in the geometrical shadow of the screen, the field U and its derivative $\frac{\partial U}{\partial n}$ are identically zero. These conditions are generally met if the opening in the screen (the aperture) is large compared to the wavelength.

1.2.2 The Fresnel–Kirchhoff diffraction formula

Another simplification can be made, if the distance r_{01} is large compared to the wavelength, the $(\mathrm{j}k - \frac{1}{r_{01}})$ term in equation $(\mathrm{j}k)$. If the aperture is illuminated by a single spherical wave at point P_2,

$$U(\mathrm{P}_1) = \frac{A\exp(\mathrm{j}kr_{21})}{r_{21}}. \tag{1.13}$$

the field at P_0 reduces to the Fresnel–Kirchhoff diffraction formula:

$$U(\mathrm{P}_0) = \frac{A}{\mathrm{j}\lambda}\iint_{\mathrm{Aper}}\left\{\frac{\exp\left[\mathrm{j}k(r_{21}+r_{01})\right]}{r_{21}r_{01}}\left[\frac{\cos(\boldsymbol{n},\boldsymbol{r}_{01}) - \cos(\boldsymbol{n},\boldsymbol{r}_{21})}{2}\right]\right\}\mathrm{d}S. \tag{1.14}$$

The integration is now over the aperture in the region around P_1. This equation is symmetrical between P_0 and P_2. The field at P_0 from the source at P_2 through the aperture would be the same as the field at P_2 from a point source at P_0. The fact that light and diffraction do not seem to care whether the beam is going left-to-right or right-to-left is known as the reciprocity theorem of Helmholtz:

$$U(\mathrm{P}_0) = \frac{A}{\mathrm{j}\lambda}\iint_{\mathrm{Aper}} U'(\mathrm{P}_1)\frac{\exp\left[\mathrm{j}k(r_{01})\right]}{r_{01}}\mathrm{d}S. \tag{1.15}$$

where

$$U'(\mathrm{P}_1) = \frac{A}{\mathrm{j}\lambda}\frac{\exp[\mathrm{j}k(r_{21})]}{r_{21}}\left[\frac{\cos(\boldsymbol{n}, \boldsymbol{r}_{01}) - \cos(\boldsymbol{n}, \boldsymbol{r}_{21})}{2}\right]. \tag{1.16}$$

The $\boldsymbol{r}$ vectors end in the aperture at P_1. As one moves across the aperture, the directions of the $\boldsymbol{r}$ vectors change and it appears that the field in the aperture is an infinite number of secondary spherical waves that match the Huygens principle. Fresnel made the Huygens assumption work along with Young's principle of interference and Kirchhoff showed that this property was a fundamental part of the nature of light itself.

1.3 The Rayleigh–Sommerfeld formulation of diffraction

Not all apertures are illuminated by a point source of light in front of the aperture. To expand to practical light sources with varying amplitudes and phases, the Rayleigh–Sommerfeld formulation is produced. Kirchhoff theory imposed two conditions on both the field amplitude and its derivative.

Going back to the Green's function solution for the field at P_0,

$$U(\mathrm{P}_0) = \frac{1}{4\pi}\iint_{\mathrm{Aper}}\left\{\frac{\partial U}{\partial n}G - U\frac{\partial G}{\partial n}\right\}\mathrm{d}S \tag{1.17}$$

where $G = [\frac{\exp(\mathrm{j}kr_{01})}{r_{01}}]$.

Sommerfeld found a Green's function that allowed either G or $\frac{\partial G}{\partial n}$ to vanish without requiring both to vanish on the aperture. He introduced another source point $\tilde{\mathrm{P}}_0$ at a mirror image of the observation point P_0. The first Rayleigh–Sommerfeld solution is

$$G_{-} = \left[\frac{\exp(\mathrm{j}kr_{01})}{r_{01}}\right] - \left[\frac{\exp\left(\mathrm{j}kr_{01}^{2}\right)}{r_{01}^{2}}\right]. \tag{1.18}$$

The new source at $\tilde{\mathrm{P}}_0$ is 180° out of phase with the point P_0 so the field vanishes on the screen and aperture.

The second Rayleigh–Sommerfeld solution is

$$G_{+} = \left[\frac{\exp(\mathrm{j}kr_{01})}{r_{01}}\right] + \left[\frac{\exp\left(\mathrm{j}kr_{01}^{2}\right)}{r_{01}^{2}}\right]. \tag{1.19}$$

The new source at P_0 is in phase with the point P_0 so the normal derivative of the field vanishes on the screen and aperture. After about two pages of mathematical rigor, we find two results for the field at P_0 due to a spherical wave source at P_2.

$$U_{\mathrm{RS1}}(\mathrm{P}_0) = \frac{A}{\mathrm{j}\lambda}\iint_{\mathrm{Aper}}\left\{\frac{\exp[\mathrm{j}k(r_{21} + r_{01})]}{r_{21}r_{01}}\cos(\boldsymbol{n}, \boldsymbol{r}_{01})\right\}\mathrm{d}S \tag{1.20}$$

$$U_{\mathrm{RS2}}(\mathrm{P}_0) = \frac{A}{\mathrm{j}\lambda}\iint_{\mathrm{Aper}} \left\{ \frac{\exp[\mathrm{j}k(r_{21} + r_{01})]}{r_{21}r_{01}} \cos(\boldsymbol{n}, \boldsymbol{r}_{21}) \right\} \mathrm{d}S \tag{1.21}$$

The Fresnel–Kirchhoff formulation, (1.14), and the two Rayleigh–Sommerfeld integrals, (1.20) and (1.21), all describe the radiation field at point P_0 as a sum of diverging spherical waves across the aperture much like Huygens' model. For cases many wavelengths from the aperture, the Fresnel–Kirchhoff solution has been shown to be the average of the two Rayleigh–Sommerfeld solutions. Only where the cosine terms (the obliquity factors) are significant near the aperture do these theories differ. The Huygens–Fresnel principle can be simplified by describing the combination of spherical waves in the aperture as a single field $U(\mathrm{P}_1)$.

References

Goodman J W 2005 *Introduction to Fourier Optics* 3rd edn (Englewood, CO: Roberts)

Kirchhoff G 1883 To the theory of light beams *Ann. Phys. Lpz.* **254** 663–95

Chapter 2

Fourier analysis and linear systems

2.1 Fourier transforms and Fourier analysis

In the 19th century, French mathematician Joseph Fourier showed how a periodic mathematical function could be decomposed into more elementary periodic mathematical functions such as trigonometric sines and cosines. With considerable mathematical rigor, the work of Fourier has been expanded into multiple dimensions and has been applied to thousands of physical problems including the field of optics.

If we start with a function $g(t)$ which is periodic, absolutely integrable over the infinite t line and a few other conditions, we can find the Fourier spectrum or frequency spectrum of the function by applying the Fourier transform defined by

$$\mathrm{FT}(g) = G(\nu) = \int_{-\infty}^{\infty} g(t)\exp[-\mathrm{j}2\pi\nu t]\mathrm{d}t. \tag{2.1}$$

Note that the function $\mathrm{FT}(g)$ will be a function of ν, the frequency variable. Although I use t and ν as the Fourier transform variable pairs, which often represent time and temporal frequency, any variable can be used. If we know the spectrum $G(\nu)$ we can find the function $g(t)$ by taking the inverse Fourier transform

$$\mathrm{FT}^{-1}[G(\nu)] = g(t) = \int_{-\infty}^{\infty} G(\nu)\exp[+\mathrm{j}2\pi\nu t]\mathrm{d}\nu. \tag{2.2}$$

2.2 Fourier analysis in two dimensions

Optical electromagnetic field propagation, such as a beam going through an aperture, is often represented by a two-dimensional wavefront propagating in a third dimension. Let us put the wavefront in the x–y plane assuming that the wave is propagating along the z-direction. We will make use of a two-dimensional Fourier

doi:10.1088/978-0-750-31056-7ch2

transform where the spatial coordinate function $g(x, y)$ is Fourier transformed into its two-dimensional spectrum with spatial frequency coordinates f_x, f_y.

$$\mathrm{FT}[g(x, y)] = G(f_x, f_y) = \int_{-\infty}^{\infty}\int g(x, y)\exp[-\mathrm{j}2\pi(f_x x + f_y y)]\,\mathrm{d}x\,\mathrm{d}y \quad (2.3)$$

and its inverse

$$\mathrm{FT}^{-1}[G(f_x, f_y)] = g(x, y) = \int\int_{-\infty}^{\infty} G(f_x, f_y)\exp[+\mathrm{j}2\pi(f_x x + f_y y)]\,\mathrm{d}f_x\,\mathrm{d}f_y. \quad (2.4)$$

The mathematical conditions of existence of Fourier transforms are very strict. In some cases, the condition of 'no infinite discontinuities' is violated by physical possibilities, such as an infinitesimally short pulse in time, or a two- or three-dimensional infinitesimal point source. In such cases, some mathematical conveniences are made.

For example, the short time pulse is given by the Dirac delta function $\delta(t)$, which is infinite at time t and zero elsewhere, but integrates over all time to the value of unity (Dirac 1958).

$$\int_{-\infty}^{\infty} \delta(t)\,\mathrm{d}t = 1 \quad (2.5)$$

Similarly, a point source is represented by a two-dimensional Dirac delta function $\delta(x, y)$, which is infinite value at $x = 0$, $y = 0$ and zero elsewhere but, integrated over all two-dimensional space, equals unity.

$$\int_{-\infty}^{\infty}\int \delta(x, y)\,\mathrm{d}x\,\mathrm{d}y = 1 \quad (2.6)$$

For mathematical convenience, but not at all a physical reality, the two-dimensional Dirac delta function point source is often represented by a limit.

$$\delta(x, y) = \lim_{N\to\infty} N^2\exp[-N^2\pi(x^2 + y^2)]. \quad (2.7)$$

The N has no physical meaning, but it allows the conditions of infinite value at (0, 0) and zero value elsewhere with the integral equal to unity. The ultimate result of the Fourier property is that we can decompose a function, whether it is one, two, three or more dimensions, into a series of functions with the specific form $\exp(\mathrm{j}2\pi\nu t)$ where the frequencies ν are just Fourier conjugates of the other dimension t. The Rayleigh–Sommerfeld equations show a form of $\exp[-\mathrm{j}2\pi(f_x x + f_y y)]$ with the spatial frequencies of the form $\frac{2\pi}{\lambda}$.

The Fourier spectrum of the function $g(x, y)$ is $G_s(f_x, f_y)$, which describes the weighting factors applied to each elementary function, or $[G(f_x, f_y)] \times \exp[-j2\pi(f_x x + f_y y)]$. These $g(x, y)$ functions are periodic and the distance between lines of zero phase is found to be

$$L = \frac{1}{\sqrt{f_x^2 + f_y^2}}. \tag{2.8}$$

2.3 Fourier transform theorems

2.3.1 Linearity

One advantage of Fourier transforms is that they are linear. That is, the transform of a sum of linear functions $g_s = g_1(x, y) + g_2(x, y) \cdots$ is the linear sum of the transforms

$$\begin{aligned} G_s(f_x, f_y) = &\int_{-\infty}^{\infty}\int g_1(x, y)\exp[-j2\pi(f_x x + f_y y)]\,dx\,dy \\ &+ \int_{-\infty}^{\infty}\int g_2(x, y)\exp[-j2\pi(f_x x + f_y y)]\,dx\,dy \cdots. \end{aligned} \tag{2.9}$$

2.3.2 Similarity

Another property is similarity. If

$$\mathrm{FT}[g(x, y)] = G(f_x, f_y) \tag{2.10}$$

then

$$\mathrm{FT}[g(ax, by)] = \frac{1}{|ab|}G\left(\frac{f_x}{a}, \frac{f_y}{b}\right). \tag{2.11}$$

A stretch in coordinate in the spatial domain is a contraction (a 'scrunch') in the dimensions in the frequency domain. Here is the first evidence that 'large' functions transform into 'small' spectra. This appears in optics as a large aperture creating a small focal spot.

2.3.3 Shifting

If a function is shifted in the spatial domain, there is a corresponding phase shift in the frequency domain.

$$\mathrm{FT}[g(x - a, y - b)] = G(f_x, f_y)\exp[-j2\pi(f_x a + f_y b)] \tag{2.12}$$

2.3.4 Parseval's theorem

If we integrate the square of the function over the entire spatial domain, we have calculated the *energy* of the function. It can be shown to be equal to the integral of the square of the spectrum.

$$\iint_{-\infty}^{\infty} |g(x, y)|^2 \, \mathrm{d}x \, \mathrm{d}y = \iint_{-\infty}^{\infty} |G(f_x, f_y)|^2 \, \mathrm{d}f_x \, \mathrm{d}f_y . \tag{2.13}$$

This formulation can be interpreted as the square of the spectrum is the energy density in the frequency domain.

2.3.5 Convolution

Convolution of two functions in the spatial domain is very important for Fourier optics and the discussion of linear effects such as image formation and spatial filtering. Two functions (of any dimensionality) are convolved when one of them has their coordinates inverted (changing signs) and shifted. Beginning at a large separation, they are dragged together. At each separation of their coordinate origins, the overlap (or product) of the two functions is calculated. Convolution of two-dimensional functions ends up with a two-dimensional function of the separation. The convolution operation is represented by ⊗.

$$g(\xi, \eta) \otimes h(\xi, \eta) = \iint_{-\infty}^{\infty} g(\xi, \eta) h(x - \xi, y - \eta) \mathrm{d}\xi \, \mathrm{d}\eta \tag{2.14}$$

The beauty of the process is that the Fourier transform of the convolution is simply the product of the Fourier transforms of each of the functions.

$$\begin{aligned} \mathrm{FT}\left[\iint_{-\infty}^{\infty} g(\xi, \eta) h(x - \xi, y - \eta) \mathrm{d}\xi \, \mathrm{d}\eta\right] &= \mathrm{FT}[g(\xi, \eta)] \, \mathrm{FT}[h(\xi, \eta)] \\ &= G(f_x, f_y) H(f_x, f_y) \end{aligned} \tag{2.15}$$

2.3.6 Autocorrelation

If the two functions are equal, that is $g = h$, a convolution becomes an auto-correlation. By convolving $g(x, y)$ with the complex conjugate of the inverted function $g^*(-x, -y)$, we get an autocorrelation. The Fourier transform of the autocorrelation is the square of the Fourier transform of the original function.

$$\mathrm{FT}\left[\iint_{-\infty}^{\infty} g(\xi, \eta) g^*(\xi - x, \eta - y) \mathrm{d}\xi \, \mathrm{d}\eta\right] = |\mathrm{FT}[g(\xi, \eta)]|^2 = |G(f_x, f_y)|^2 \tag{2.16}$$

or

$$\mathrm{FT}[g(x, y)^2] = \left[\iint_{-\infty}^{\infty} G(\xi, \eta) G^*(\xi - f_x, \eta - f_y) \mathrm{d}\xi \, \mathrm{d}\eta\right]. \tag{2.17}$$

Detailed descriptions of many optical-related properties of convolution can be found elsewhere (Gaskill 1978, Papoulis 1962).

2.3.7 Inverse Fourier transform

For each point in the continuous function, the transform of the inverse transform equals the inverse transform of the transform which equals the function. Mathematics is such an amazing tool that it can translate this incomprehensible sentence into

$$\mathrm{FT}\,\mathrm{FT}^{-1}[g(x, y)] = \mathrm{FT}^{-1}\,\mathrm{FT}[g(x, y)] = g(x, y). \tag{2.18}$$

2.4 Separable functions

In general, two-dimensional functions and two-dimensional transforms can be unwieldy. Where some functions are only finite over certain regions or some clearly do not indicate periodicity, Fourier transforms are not the best tools. However, in a lot of physical systems, these difficulties disappear and can even become simpler than the two-dimensional integrals we have been discussing.

One of the primary simplifications occurs when the two-dimensional function $g(x, y)$ is separable into two functions, one solely in x and one in y. These coordinates do not have to be Cartesian (x, y), but they can be any two orthogonal coordinates, such as polar coordinates (r, θ). When this separability occurs we can reduce the two-dimensional manipulations to one-dimensional manipulations.

$$\begin{aligned} \mathrm{FT}[g(x, y)] &= \int_{-\infty}^{\infty}\int g(x, y)\exp[-\mathrm{j}2\pi(f_x x + f_y y)]\,\mathrm{d}x\,\mathrm{d}y \\ &= \int_{-\infty}^{\infty} g_x(x)\exp[-\mathrm{j}2\pi(f_x x)]\,\mathrm{d}x \int_{-\infty}^{\infty} g_x(y)\exp[-\mathrm{j}2\pi(f_y y)]\,\mathrm{d}y \\ &= \mathrm{FT}_x\{g_x\}(f_x)\ \mathrm{FT}_x\{g_x\}(f_y) \end{aligned} \tag{2.19}$$

2.4.1 Functions with circular symmetry: Fourier–Bessel transforms

One set of conditions appears in optics countless times. The iris in the human eyeball produces a circular pupil for the eye. Millions of years of evolution have found that circular symmetry in problems dealing with light is a good thing. Without too much more philosophizing, the functions, be they beams going through a circular aperture, the refraction and transmission through a circular lens or a truncated Gaussian laser beam, all have circular symmetry. That means that there is no dependence on the azimuthal variable θ.

$$g(r, \theta) = g_R(r) \tag{2.20}$$

The Fourier transform in Cartesian coordinates

$$G(f_x, f_y) = \int_{-\infty}^{\infty}\int g(x, y)\exp[-\mathrm{j}2\pi(f_x x + f_y y)]\,\mathrm{d}x\,\mathrm{d}y \tag{2.21}$$

can be translated into polar coordinates with a few identities. The coordinates transform as

$$r = \sqrt{x^2 + y^2} \qquad x = r\cos\theta$$
$$\theta = \arctan\frac{y}{x} \qquad y = r\sin\theta. \tag{2.22}$$

The spatial frequency coordinates transform as

$$\rho = \sqrt{f_x^{\,2} + f_y^{\,2}} \qquad f_x = \rho\cos\phi$$
$$\phi = \arctan\frac{f_y}{f_x} \qquad f_y = \rho\sin\phi. \tag{2.23}$$

Applying the coordinate transformations to (2.21)

$$G(\rho, \phi) = \int_0^{2\pi} \exp[-j2\pi r\rho\cos(\theta - \phi)]\,d\theta \int_0^{\infty} g_R(r)\, r\, dr. \tag{2.24}$$

Now, using a Bessel function identity found in most calculus texts (Abramowitz and Stegun 1972),

$$J_0(a) = \frac{1}{2\pi}\int_0^{2\pi} \exp[-ja\cos(\theta - \phi)]\,d\theta \tag{2.25}$$

where $J_0(\,)$ is a zero-order (notice the subscript) Bessel function of the first-kind (its symbol is J for historical reasons). Notice how cleverly the function of ϕ disappeared. The Fourier transform now reduces to a one-dimensional function $G_o(\rho)$

$$G_o(\rho, \phi) = G_o(\rho) = 2\pi\int_0^{\infty} g_R(r)\, J_0(2\pi\rho)\, r\, dr \tag{2.26}$$

where the lowercase 'o' subscript to G represents the circular symmetry. The new transform is called a *Hankel transform of zero order* or a *Fourier–Bessel transform*, symbolized by FB. The previous properties of Fourier transforms hold.

$$\mathrm{FB}\,\mathrm{FB}^{-1}[g_R(r)] = \mathrm{FB}^{-1}\,\mathrm{FB}[g_R(r)] = \mathrm{FB}\,\mathrm{FB}[g_R(r)] = g(x, y) \tag{2.27}$$

The third term indicates that the Fourier–Bessel transform has symmetry and the integral is from the origin to infinity.

The similarity theorem also holds. That is, a scaled function has a scaled transform.

$$\mathrm{FB}[g_R(ar)] = \frac{1}{a^2}G_o\left(\frac{\rho}{a}\right) \tag{2.28}$$

Throughout all these discussions, one should keep in mind the physicality of the equations. A two-dimensional spatial Fourier transform or a Fourier–Bessel transform has the units of the original function *times* area. The units of spatial frequency are always inverse distance. For example, if the units of the original function are radiometric watts per square meter, the Fourier transform is in watts as is the simple two-dimensional integral of the function. Consistency of units is a good way to check one's first calculations to avoid the common errors of misinterpreting the meaning of what the mathematics is telling us.

2.5 Useful functions and Fourier transform pairs

In the process of working with Fourier optics, it is necessary to use a number of specialized and specially named mathematical functions. These are defined here.

A *rectangle function* is a one-dimensional rectangle of width 1 unit, defined as

$$\mathrm{rect}(x) = \begin{cases} 1 & |x| < \dfrac{1}{2} \\ \dfrac{1}{2} & |x| = \dfrac{1}{2} \\ 0 & \text{otherwise.} \end{cases} \tag{2.29}$$

It is important to include the value of $\frac{1}{2}$ on the discontinuity at $|x| = \frac{1}{2}$.

A *signum function* carries the sign of a variable.

$$\mathrm{sgn}(x) = \begin{cases} 1 & x > 0 \\ 0 & x = 0 \\ -1 & x < 0 \end{cases} \tag{2.30}$$

A *sinc* function (pronounced 'sink') is the sine of a variable divided by the variable.

$$\mathrm{sinc}(x) = \frac{\sin(\pi x)}{\pi x} \tag{2.31}$$

A *triangle* function is a triangle!

$$\Lambda(x) = \begin{cases} 1 - |x| & |x| \leqslant 1 \\ 0 & \text{otherwise} \end{cases} \tag{2.32}$$

A Dirac delta function is represented by a vertical arrow pointing upward. It is infinitesimally thin and infinitely high. Its area is equal to unity. An infinite series of delta functions appears to be a comb, thus it is called the *comb* function.

$$\mathrm{comb}(x) = \sum_{n=-\infty}^{\infty} \delta(x - n) \tag{2.33}$$

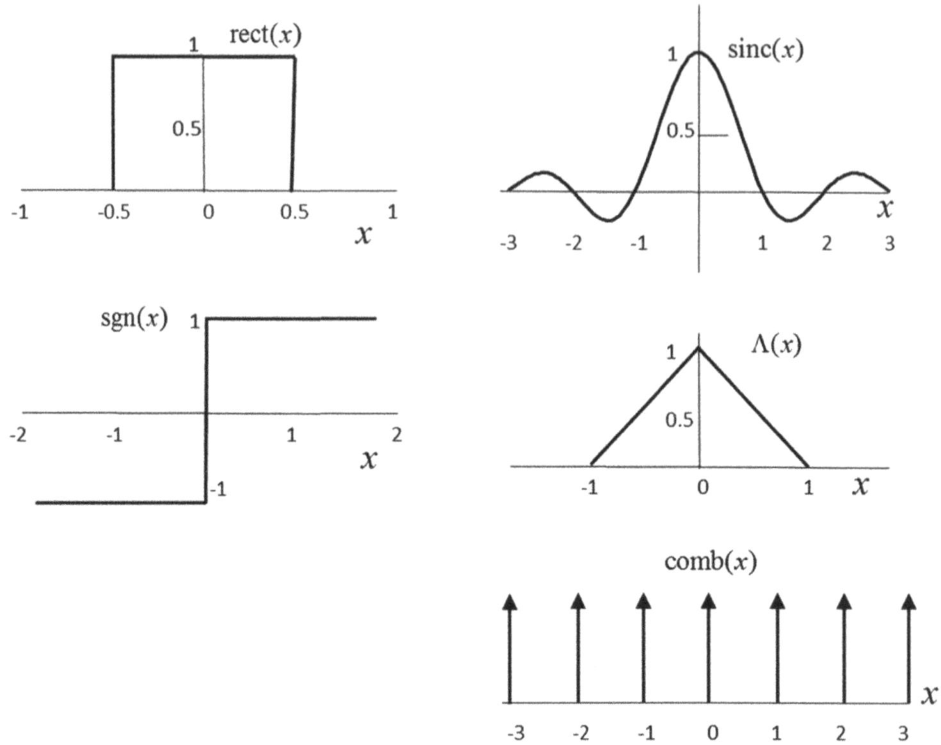

Figure 2.1. One-dimensional special functions.

A fundamental two-dimensional function (especially in optics) is the *circle* function. It is unity inside the circle and zero outside the circle. Note again that its value on the circle is $\frac{1}{2}$.

$$\mathrm{circ}\left(\sqrt{x^2+y^2}\right) = \begin{cases} 1 & \sqrt{x^2+y^2} < 1 \\ \dfrac{1}{2} & \sqrt{x^2+y^2} = 1 \\ 0 & \text{otherwise.} \end{cases} \tag{2.34}$$

In polar coordinates, the circle function is

$$\mathrm{circ}(r) = \begin{cases} 1 & r < 1 \\ \dfrac{1}{2} & r = 1 \\ 0 & \text{otherwise.} \end{cases} \tag{2.35}$$

See figure 2.1.

Many separable functions can be Fourier transformed using the one-dimensional functions described above. Table 2.1 shows a number of these transform pairs.

Table 2.1. Fourier transform pairs.

Function	Transform
$\exp[-\pi(a^2x^2 + b^2y^2)]$	$\frac{1}{\lvert ab\rvert}\exp\left[-\pi\left(\frac{f_x^2}{a^2} + \frac{f_y^2}{b^2}\right)\right]$
$\mathrm{rect}(ax)\mathrm{rect}(by)$	$\frac{1}{\lvert ab\rvert}\mathrm{sinc}\left(\frac{f_x}{a}\right)\mathrm{sinc}\left(\frac{f_y}{b}\right)$
$\Lambda(ax)\Lambda(by)$	$\frac{1}{\lvert ab\rvert}\mathrm{sinc}^2\left(\frac{f_x}{a}\right)\mathrm{sinc}^2\left(\frac{f_y}{b}\right)$
$\delta(ax, by)$	$\frac{1}{\lvert ab\rvert}$
$\exp[\mathrm{j}\pi(ax + by)]$	$\delta\left(f_x - \frac{a}{2}, f_y - \frac{b}{2}\right)$
$\mathrm{sgn}(ax)\,\mathrm{sgn}(by)$	$\frac{ab}{\lvert ab\rvert}\frac{1}{\mathrm{j}\pi f_x}\frac{1}{\mathrm{j}\pi f_y}$
$\mathrm{comb}(ax)\mathrm{comb}(by)$	$\frac{1}{\lvert ab\rvert}\mathrm{comb}\left(\frac{f_x}{a}\right)\mathrm{comb}\left(\frac{f_y}{b}\right)$
$\exp[-\mathrm{j}\pi(a^2x^2 + b^2y^2)]$	$\frac{\mathrm{j}}{\lvert ab\rvert}\exp\left[-\mathrm{j}\pi\left(\frac{f_x^2}{a^2} + \frac{f_y^2}{b^2}\right)\right]$
$\exp[-(a\lvert x\rvert + b\lvert y\rvert)]$	$\frac{1}{\lvert ab\rvert}\frac{2}{1+\left(\frac{2\pi f_x}{a}\right)^2}\frac{2}{1+\left(\frac{2\pi f_y}{b}\right)^2}$
$\mathrm{FB}\{\mathrm{circ}(r)\}$	$\frac{J_1(2\pi\rho)}{\rho}$

The Fourier–Bessel transform of the polar circ function, $\frac{J_1(2\pi\rho)}{\rho}$, has a value of π at the origin. See figure 2.2. When this function is normalized, the value is unity at the origin.

$$\frac{2J_1(2\pi\rho)}{2\pi\rho} \tag{2.36}$$

This function is called a '*jinc*' function; 'j' as in the Bessel symbol, 'inc' because it follows the form of the sinc function. Scientists at the University of Arizona recognized the shape of the function as it resembled a Mexican sombrero. It is also called a sombrero function, somb(2ρ) (Gaskill 1978). You can't make this stuff up! Well, yes you can.

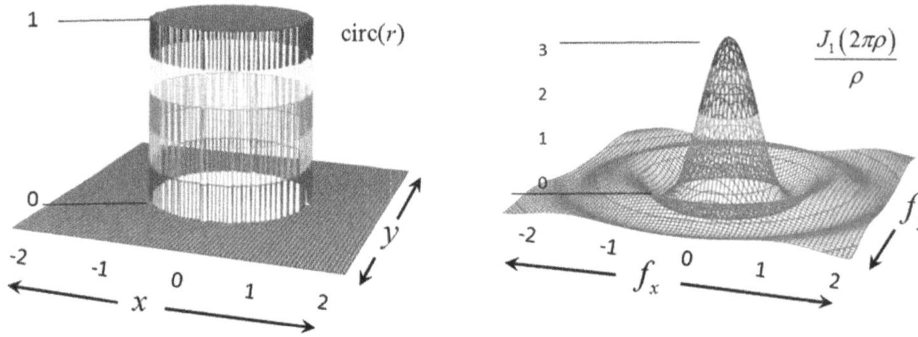

Figure 2.2. The circle function and its transform, the *jinc* or sombrero function.

2.6 Linear systems

2.6.1 Linearity and the superposition integral

Optical beam propagation has been shown to be a compilation of a series of beams propagating from points on the wavefront (the Huygens principle). Imaging systems can be depicted as optical beams propagating from points on the object through an imaging system to form an image made up of an intensity map over space. In these optical systems, the functions can be real-valued, such as optical intensity, or complex-valued, as in electromagnetic field amplitudes.

It is our intention to determine how we map an optical input to an output, or an object to an image. We can use an operator notation, with the operator to be determined later, that maps the input two-dimensional function $g_1(x_1, y_1)$ to the output function $g_2(x_2, y_2)$. The subscripts generally indicate the position along some third axis, such that these two functions correspond to an input plane 1 and an output plane 2. Using the notation of $S(\,)$ for the operator, the input-to-output process is given by

$$g_2(x_2, y_2) = S[g_1(x_1, y_1)]. \tag{2.37}$$

For our purposes here, we will assume that the operator is linear. That is

$$S[ag_1'(x_1, y_1) + bg_1''(x_1, y_1)] = aS[g_1'(x_1, y_1)] + bS[g_1''(x_1, y_1)] \tag{2.38}$$

where $g_1'(x_1, y_1)$ and $g_1''(x_1, y_1)$ are the input functions and a and b can be complex constants.

One interesting consequence of Dirac delta functions is their effect on integration. Because a function has *any* value *only* at its origin (0, 0), it acts as an *impulse*. It will only have a value whenever the variables in the coordinate argument of $\delta(x, y)$

are zero. So, the integral of a function multiplied by a delta function can only take on one specific value.

$$\iint_{-\infty}^{\infty} g_1(\xi, \eta)\, \delta(x_1 - \xi, y_1 - \eta)\, \mathrm{d}\xi\, \mathrm{d}\eta = g_1(x_1, y_1) \tag{2.39}$$

Therefore, while the function $g_1(x, y)$ has values all along the coordinates, the integration of the product with the delta function pulls out only one particular value, the value at coordinates (x_1, y_1). The function $g_1(x, y)$ is a linear combination of weighted and displaced delta functions $\delta(x_1 - \xi, y_1 - \eta)$. The weights are $g_1(\xi, \eta)$. The mathematics has been compared to a geologist's or chef's sifting screen that intentionally allows only certain values of the material (the small pieces) to filter through. This property of the delta function is called the *sifting property* and it will allow us to use the operators and their linearity property to sift through the elementary functions (the point sources along the wavefront) that can make up a complicated optical signal.

We can now calculate the output distribution by using the operator on $g_1(x, y)$.

$$g_2(x_2, y_2) = S\left\{\iint_{-\infty}^{\infty} g_1(\xi, \eta)\, \delta(x_1 - \xi, y_1 - \eta)\, \mathrm{d}\xi\, \mathrm{d}\eta\right\} \tag{2.40}$$

Because the operator is linear and operates on each value independently, we bring the operator inside the integral

$$g_2(x_2, y_2) = \iint_{-\infty}^{\infty} g_1(\xi, \eta)\, S[\delta(x_1 - \xi, y_1 - \eta)]\, \mathrm{d}\xi\, \mathrm{d}\eta. \tag{2.41}$$

The output function is seen as the weighted accumulation of individual responses to the operator for simple input values, a *superposition* of the point sources or delta functions. In optics, the response of an imaging system to a point source is called the *point spread function* (PSF). Thus we define the PSF as

$$h(x_2, y_2, \xi, \eta) = S[\delta(x_1 - \xi, y_1 - \eta)]\,. \tag{2.42}$$

2.6.2 Invariant linear systems and transfer functions

Imaging systems are considered space-invariant when the impulse response (the PSF) depends only on the distances between the object point and the image points, $(x_2 - \xi)$ and $(y_2 - \eta)$.

$$h(x_2, y_2, \xi, \eta) = h(x_2 - \xi, y_2 - \eta) \tag{2.43}$$

The image of a point source changes only in location, not in its functional form $h(\,)$. If this occurs precisely over the entire plane of the image, the region is called isoplanatic. While often this is not exactly true, especially when aberrations occur, the imaging process is considered to be linear and space-invariant in the isoplanatic region near the optical axis.

The image,

$$g_2(x_2, y_2) = \iint_{-\infty}^{\infty} g_1(\xi, \eta)\, h(x_1 - \xi, y_1 - \eta)\, \mathrm{d}\xi\, \mathrm{d}\eta \tag{2.44}$$

is now seen to be a convolution of the object response $g_1(\xi, \eta)$ with the PSF of the system. In shorthand,

$$g_2 = g_1 \otimes h. \tag{2.45}$$

Now, by Fourier transforming both sides and using the convolution property,

$$G_2(f_x, f_y) = H(f_x, f_y) G_1(f_x, f_y). \tag{2.46}$$

The imaging process becomes a simple product of two functions, the Fourier transform of the object distribution (the object spectrum) and the Fourier transform of the PSF (the transfer function of the system).

$$H(f_x, f_y) = \int_{-\infty}^{\infty}\!\!\int h(\xi, \eta) \exp[-\mathrm{j}2\pi(f_x\xi + f_y\eta)]\, \mathrm{d}\xi\, \mathrm{d}\eta. \tag{2.47}$$

The image distribution is found by inverse transforming the image spectrum.

References

Abramowitz M and Stegun I A 1972 *Handbook of Mathematical Functions* (New York: Dover)
Dirac P 1958 *Principles of Quantum Mechanics* 4th edn (Oxford: Clarendon)
Gaskill J D 1978 *Linear Systems, Fourier Transforms, and Optics* (New York: Wiley)
Papoulis A 1962 *The Fourier Integral and Its Applications* (New York: McGraw-Hill)

Chapter 3

Fresnel and Fraunhofer diffraction and wave optics

Continuing from chapter 1 and using the linear formulation from chapter 2, we can rewrite the Rayleigh–Sommerfeld solution $U(P_0)$ as a description of a superposition of spherical waves $\frac{\exp jkr_{01}}{r_{01}}$ coming from secondary Huygens sources over the aperture $U(P_1)$.

$$U(P_0) = \frac{1}{j\lambda}\iint_{\text{Aper}} U(P_1)\frac{\exp jkr_{01}}{r_{01}}\cos\theta \, dS \tag{3.1}$$

The angle θ is the angle between the vector $\boldsymbol{r}_{01}$ and the vector normal to the aperture is $\boldsymbol{n}$. $U(P)$ are complex amplitudes; the amplitude is inversely proportional to the wavelength and therefore proportional to the optical frequency ν, and a few other properties regarding phase and directivity. Now, rewriting further, we can show the diffraction integral to be

$$U(P_0) = \int_{\text{Aper}}\int h(P_0, P_1)U(P_1)\, dS \tag{3.2}$$

where the impulse response is

$$h(P_0, P_1) = \frac{\exp jkr_{01}}{r_{01}}\cos\theta. \tag{3.3}$$

3.1 Diffraction at the boundary of an aperture

By representing the complex fields $U(P)$ in terms of their Fourier transforms, or spectra, the superposition of spherical waves can be decomposed into a superposition of plane waves traveling in various directions through any one plane. We begin by assuming a monochromatic wave is traveling in the z-direction. The plane wave is represented by a complex field in the x–y plane. At $z = 0$, the field is $U(x, y, 0)$. We now wish to find the field at some other plane z. The spectrum of the field $U(x, y, 0)$ is

$$A(f_x, f_y; 0) = \int_{-\infty}^{\infty}\int U(x, y, 0)\exp[-\mathrm{j}2\pi(f_x x + f_y y)]\,\mathrm{d}x\,\mathrm{d}y. \tag{3.4}$$

If we represent the spatial frequencies evaluated at unitless specific coordinates α and β, ($\alpha = f_x\lambda$, $\beta = f_y\lambda$), the result

$$A\left(\frac{\alpha}{\lambda}, \frac{\beta}{\lambda}; 0\right) = \int_{-\infty}^{\infty}\int U(x, y, 0)\exp\left[-\mathrm{j}2\pi\left(\frac{\alpha}{\lambda}x + \frac{\beta}{\lambda}y\right)\right]\mathrm{d}x\,\mathrm{d}y \tag{3.5}$$

is called the angular spectrum of the field $U(x, y, 0)$. We now want to calculate the field at some plane z. The spectrum at z is

$$A\left(\frac{\alpha}{\lambda}, \frac{\beta}{\lambda}; z\right) = \int_{-\infty}^{\infty}\int U(x, y, z)\exp\left[-\mathrm{j}2\pi\left(\frac{\alpha}{\lambda}x + \frac{\beta}{\lambda}y\right)\right]\mathrm{d}x\,\mathrm{d}y. \tag{3.6}$$

Performing the inverse transform,

$$U(x, y, z) = \int_{-\infty}^{\infty}\int A\left(\frac{\alpha}{\lambda}, \frac{\beta}{\lambda}; z\right)\exp\left[+\mathrm{j}2\pi\left(\frac{\alpha}{\lambda}x + \frac{\beta}{\lambda}y\right)\right]\mathrm{d}\frac{\alpha}{\lambda}\,\mathrm{d}\frac{\beta}{\lambda} \tag{3.7}$$

which, in addition to $U(x, y, 0)$, must also satisfy the Helmholtz equation,

$$\nabla^2 U + k^2 U = 0. \tag{3.8}$$

With direct differentiation of $U(x, y, z)$, we find

$$A\left(\frac{\alpha}{\lambda}, \frac{\beta}{\lambda}; z\right) = A\left(\frac{\alpha}{\lambda}, \frac{\beta}{\lambda}; 0\right)\exp\left(\mathrm{j}\frac{2\pi}{\lambda}\sqrt{1 - \alpha^2 - \beta^2}\, z\right). \tag{3.9}$$

The angular spectrum transforms over the coordinate system (α, β), which physically represents the direction of the plane waves that make up the field. When $\alpha^2 + \beta^2 < 1$, the argument in the exponent remains imaginary and it represents a spectrum of plane waves. When $\alpha^2 + \beta^2 > 1$

$$A\left(\frac{\alpha}{\lambda}, \frac{\beta}{\lambda}; z\right) = A\left(\frac{\alpha}{\lambda}, \frac{\beta}{\lambda}; 0\right)\exp(-\mu z). \tag{3.10}$$

Because μ is a positive real number, the spectrum decays and does not propagate very far. These so-called *evanescent waves* carry no energy away from the aperture.

3.1.1 Limiting evanescent waves

We can limit the effect of evanescent waves by multiplying the angular spectrum by an appropriate circ( ) function prior to taking the inverse transform.

$$U(x, y, z) = \int_{-\infty}^{\infty}\int A\left(\frac{\alpha}{\lambda}, \frac{\beta}{\lambda}; z\right)\exp\left(\mathrm{j}\frac{2\pi}{\lambda}\sqrt{1 - \alpha^2 - \beta^2}\, z\right)\mathrm{circ}\left(\sqrt{\alpha^2 + \beta^2}\right)$$
$$\times\exp\left[\mathrm{j}2\pi\left(\frac{\alpha}{\lambda}x + \frac{\beta}{\lambda}y\right)\right]\mathrm{d}\frac{\alpha}{\lambda}\,\mathrm{d}\frac{\beta}{\lambda}. \tag{3.11}$$

This filtering out of the evanescent waves is equivalent to showing that conventional imaging cannot resolve details smaller than the wavelength.

3.1.2 Effects of an aperture on the angular spectrum

The circ( ) function applied to the spatial frequencies limits the integration of the inverse. In each of the other spatial integrals above, we are integrating over infinite space to determine the spectra. We can limit the integration to a finite aperture at the $z = 0$ plane by creating an amplitude transmittance function in that plane. This function is simply a ratio of how much of the field is transmitted through the aperture versus how much is incident upon the aperture. It can be a complex quantity indicating phase changes.

$$t_{\mathrm{A}}(x, y) = \frac{U_t(x, y; 0)}{U_i(x, y; 0)} \tag{3.12}$$

or

$$U_t(x, y; 0) = t_{\mathrm{A}}(x, y)\, U_i(x, y; 0) \tag{3.13}$$

Using convolution, we can determine the angular spectrum of the transmitted field.

$$A_t\left(\frac{\alpha}{\lambda}, \frac{\beta}{\lambda}\right) = A_i\left(\frac{\alpha}{\lambda}, \frac{\beta}{\lambda}\right) \otimes T\left(\frac{\alpha}{\lambda}, \frac{\beta}{\lambda}\right) \tag{3.14}$$

Where $T(\frac{\alpha}{\lambda}, \frac{\beta}{\lambda})$ is the Fourier transform of $t_{\mathrm{A}}(x, y)$. For the special case of an aperture, the most common case of the aperture being a hole in an opaque screen, the incident angular spectrum is

$$A_t\left(\frac{\alpha}{\lambda}, \frac{\beta}{\lambda}\right) = \delta\left(\frac{\alpha}{\lambda}, \frac{\beta}{\lambda}\right) \tag{3.15}$$

and

$$A_t\left(\frac{\alpha}{\lambda}, \frac{\beta}{\lambda}\right) = \delta\left(\frac{\alpha}{\lambda}, \frac{\beta}{\lambda}\right) \otimes T\left(\frac{\alpha}{\lambda}, \frac{\beta}{\lambda}\right) = T\left(\frac{\alpha}{\lambda}, \frac{\beta}{\lambda}\right) \tag{3.16}$$

which is a very important result. The transmitted angular spectrum is simply the Fourier transform of the amplitude transmittance function of the aperture. The diffracting structure limits the angular frequencies in the spectrum. A small hole produces a wide range of spatial frequencies and a large hole produces a smaller range. Recall: large aperture = fine detail = high resolution.

From two expressions for the field in the z-plane,

$$\begin{aligned} U(x, y, z) = \int_{-\infty}^{\infty}\int A(f_x, f_y; 0) \exp\left(\mathrm{j}\frac{2\pi}{\lambda}\sqrt{1 - (\lambda f_x)^2 - (\lambda f_y)^2}\, z\right) \\ \times \operatorname{circ}\left(\sqrt{(\lambda f_x)^2 + (\lambda f_y)^2}\right) \exp[\mathrm{j}2\pi(f_x x + f_y y)]\, \mathrm{d}f_x\, \mathrm{d}f_y \end{aligned} \tag{3.17}$$

and

$$U(x, y, z) = \int_{-\infty}^{\infty}\int A(f_x, f_y; z)\exp[\mathrm{j}2\pi(f_x x + f_y y)]\,\mathrm{d}f_x\,\mathrm{d}f_y, \tag{3.18}$$

the angular spectrum at plane z is

$$A(f_x, f_y; z) = A(f_x, f_y; 0)\,\mathrm{circ}\left(\sqrt{(\lambda f_x)^2 + (\lambda f_y)^2}\right)\exp\left[\mathrm{j}2\pi\frac{z}{\lambda}\sqrt{1 - (\lambda f_x)^2 - (\lambda f_y)^2}\right] \tag{3.19}$$

which implies that the transfer function for the wave propagation is

$$H(f_x, f_y) = \begin{cases}\exp\left[\mathrm{j}2\pi\frac{z}{\lambda}\sqrt{1 - (\lambda f_x)^2 - (\lambda f_y)^2}\right] & \sqrt{f_x^2 + f_y^2} < \frac{1}{\lambda} \\ 0 & \text{otherwise.}\end{cases} \tag{3.20}$$

This result has been proven to show the equivalence of the Rayleigh–Sommerfeld solution and the angular spectrum approach to calculating the diffracted fields (Sherman 1967).

3.1.3 Nonmonochromatic waves

To generalize the Rayleigh–Sommerfeld result to nonmonochromatic waves, it is recognized that the expressions must be frequency dependent and therefore time dependent (Goodman 2005). An observed field behind an aperture $U(x_0, y_0, z_0, t)$ is related to the first time derivative of the field on the aperture $U(x_1, y_1, z_1, t)$ with a suitable time delay to account for the distance traveled. Where the velocity of light in the material is $v = \frac{c}{n}$, the vector from the aperture to the observation point (x_0, y_0, z_0) is $\boldsymbol{r}_{01}$ and $\boldsymbol{n}$ is the vector normal to the aperture, the relevant expression becomes

$$U(x_0, y_0, z_0, t) = \int_{\mathrm{Aper}}\int \frac{\cos(\boldsymbol{n}, \boldsymbol{r}_{01})}{2\pi v r_{01}}\frac{\mathrm{d}U\left(x_1, y_1, z_1, t - \frac{r_{01}}{v}\right)}{\mathrm{d}t}. \tag{3.21}$$

While this is considerably more tedious than purely monochromatic beam propagation, often the source has a narrow band and the monochromatic approximation can be used.

3.2 The Fresnel approximation

The Huygens–Fresnel principle as predicted by the Rayleigh–Sommerfeld solution is repeated here.

$$U(\mathrm{P}_0) = \frac{1}{\mathrm{j}\lambda}\iint_{\mathrm{Aper}} U(\mathrm{P}_1)\frac{\exp \mathrm{j}kr_{01}}{r_{01}}\cos\theta\,\mathrm{d}S \tag{3.22}$$

If we make the one assumption that $r_{01} \gg \lambda$, the cosine term is defined as $\cos\theta = \frac{z}{r_{01}}$. Assuming that point P_0 is in the x–y plane, a distance z from the ξ–η plane of P_1, we can rewrite the Huygens–Fresnel integral as

$$U(x, y) = \frac{z}{j\lambda} \iint_{\text{Aper}} U(\xi, \eta) \frac{\exp(jkr_{01})}{r_{01}^2} d\xi \, d\eta \tag{3.23}$$

where the distance is

$$r_{01} = \sqrt{z^2 + (x - \xi)^2 + (y - \eta)^2}. \tag{3.24}$$

We are now at a point where it is nearly impossible to put an arbitrary (although well-defined) field $U(\xi, \eta)$ and an arbitrary (although well-mannered) aperture geometry into (3.23) and find a closed-form solution. To obtain a reasonable and physical mathematical result, a few approximations are in order. By performing a binomial expansion of (3.24) and keeping the first two terms, we approximate the distance r_{01} as

$$r_{01} \approx z\left[1 + \frac{1}{2}\left(\frac{(x - \xi)}{z}\right)^2 + \frac{1}{2}\left(\frac{(y - \eta)}{z}\right)^2\right]. \tag{3.25}$$

$$U(x, y) = \frac{e^{jkz}}{j\lambda z} \iint_{\text{Aper}} U(\xi, \eta) \exp\left\{j\frac{k}{2z}\left[(x - \xi)^2 + (y - \eta)^2\right]\right\} d\xi \, d\eta. \tag{3.26}$$

By incorporating the aperture boundary conditions into $U(\xi, \eta)$, we get the form of a convolution

$$U(x, y) = \frac{e^{jkz}}{j\lambda z} \int_{-\infty}^{\infty}\int U_{\text{Aper}}(\xi, \eta) \exp\left\{j\frac{k}{2z}\left[(x - \xi)^2 + (y - \eta)^2\right]\right\} d\xi \, d\eta \tag{3.27}$$

whereby

$$U(x, y) = \int_{-\infty}^{\infty}\int U(\xi, \eta)\, h(x - \xi, y - \eta) d\xi \, d\eta \tag{3.28}$$

with

$$h(x, y) = \frac{e^{jkz}}{j\lambda z} \exp\left[j\frac{k}{2z}(x^2 + y^2)\right]. \tag{3.29}$$

If we factor part of the exponential (that is independent of the variables ξ or η) outside the integral we have

$$U(x, y) = \frac{e^{jkz}}{j\lambda z} \exp\left[j\frac{k}{2z}(x^2 + y^2)\right] \int_{-\infty}^{\infty}\int \left\{U_{\text{Aper}}(\xi, \eta) \exp\left[j\frac{k}{2z}(\xi^2 + \eta^2)\right]\right\}$$
$$\times \exp\left[-j\frac{2\pi}{\lambda z}(x\xi + y\eta)\right] d\xi \, d\eta \tag{3.30}$$

which is the Fourier transform of the complex field just exiting the aperture and a quadratic phase exponential resembling a focus term. Equation (3.30) is called the *Fresnel diffraction integral*, which is valid just past the aperture as long as the approximation (3.25) holds.

3.2.1 When does the approximation hold?

If we assume that the error between the Rayleigh–Sommerfeld solution and the Fresnel approximation is a phase change of a lot less than 1 radian, the following condition results,

$$z \gg \left\{ \frac{\pi}{4\lambda} \left[(x - \xi)^2 + (y - \eta)^2 \right]^2_{\max} \right\}^{\frac{1}{3}}. \tag{3.31}$$

In many cases, this condition is much too strict because much of the contribution to the Fresnel integral comes from the points that are near normal to the aperture plane, that is, $x \approx \xi$ and $y \approx \eta$.

Another way to look at the Fresnel approximation is to compare it with the transfer function of free-space propagation, (3.20). Fourier transforming (3.29) becomes

$$H(f_x, f_y) = \mathrm{FT}\left\{ \frac{\mathrm{e}^{\mathrm{j}kz}}{\mathrm{j}\lambda z} \exp\left[\mathrm{j}\frac{k}{2z}(x^2 + y^2) \right] \right\} = \mathrm{e}^{\mathrm{j}kz} \exp\left[-\mathrm{j}\pi\lambda z\left(f_x^2 + f_y^2\right) \right] \tag{3.32}$$

which implies the binomial expansion

$$\sqrt{1 - (\lambda f_x)^2 - (\lambda f_y)^2} \approx 1 - \frac{(\lambda f_x)^2}{2} - \frac{(\lambda f_y)^2}{2} \tag{3.33}$$

which restricts the results to small angles, i.e. $x \approx \xi$ and $y \approx \eta$.

3.2.2 Intensity as the physically measured quantity

Semiconductor photon detectors do not detect wave fields $U(\ )$ with amplitudes and phases, but rather they respond to optical power P_{Power}, which generates electrons and holes in the semiconductor material and a corresponding photocurrent i that is usually proportional to the power.

$$i = \Re P_{\mathrm{Power}} \tag{3.34}$$

The responsivity of the detector $\Re$ is related to the quantum efficiency of the material and other factors. The total power on a physically finite detector is the integral of the power density $p(x, y)$ over the detector area.

$$P_{\mathrm{Power}} = \iint_{\mathrm{Det.Area}} p(x, y)\,\mathrm{d}x\,\mathrm{d}y \tag{3.35}$$

The power density can be derived from the wave electric field vectors.

$$p(x, y) = \frac{\boldsymbol{E}_0 \ \boldsymbol{E}_0^*}{2\sqrt{\frac{\mu}{\varepsilon}}} \tag{3.36}$$

is directly proportional to the optical intensity $I(\mathrm{P}_0)$. This intensity (of the scalar monochromatic wave) is the squared magnitude of the field $U(\mathrm{P}_0)$.

$$I(\mathrm{P}_0) = \left|U(\mathrm{P}_0)\right|^2 \tag{3.37}$$

3.3 Fresnel diffraction calculations

3.3.1 Fresnel diffraction by a square aperture

For a very few aperture geometries, we can calculate the Fresnel diffraction pattern in closed form. As an example, consider the diffraction through a square aperture of width $2a$. The complex field directly after the aperture is

$$U(\xi, \eta) = \mathrm{rect}\left(\frac{\xi}{2a}\right)\mathrm{rect}\left(\frac{\eta}{2a}\right). \tag{3.38}$$

Repeating (3.27) with the integral limits indicating the aperture,

$$U(x, y) = \frac{\mathrm{e}^{\mathrm{j}kz}}{\mathrm{j}\lambda z}\int_{-a}^{a}\int U(\xi, \eta)\exp\left\{\mathrm{j}\frac{k}{2z}\left[(x-\xi)^2 + (y-\eta)^2\right]\right\}\mathrm{d}\xi\,\mathrm{d}\eta. \tag{3.39}$$

The integral can be separated into two one-dimensional integrals:

$$U(x) = \frac{\mathrm{e}^{\mathrm{j}kz}}{\mathrm{j}\lambda z}\int_{-a}^{a}\exp\left[\mathrm{j}\frac{k}{2z}(\xi - x)^2\right]\mathrm{d}\xi \tag{3.40}$$

and

$$U(y) = \frac{\mathrm{e}^{\mathrm{j}kz}}{\mathrm{j}\lambda z}\int_{-a}^{a}\exp\left[\mathrm{j}\frac{k}{2z}(\eta - y)^2\right]\mathrm{d}\eta. \tag{3.41}$$

With a few definitions and a change of variables for the limits of integration, we can find the expression for the field $U(\ )$ as a function of z.

$$\begin{aligned} U(x, y) = \frac{\mathrm{e}^{\mathrm{j}kz}}{2\mathrm{j}}&\left\{\left[C(\alpha_2) - C(\alpha_1)\right] + \mathrm{j}\left[S(\alpha_2) - S(\alpha_1)\right]\right\} \\ &\times\left\{\left[C(\beta_2) - C(\beta_1)\right] + \mathrm{j}\left[S(\beta_2) - S(\beta_1)\right]\right\} \end{aligned} \tag{3.42}$$

The variables are defined as

$$\begin{aligned} \alpha_1 &= -\sqrt{\frac{2}{\lambda z}}(a + x) \qquad & \alpha_2 &= +\sqrt{\frac{2}{\lambda z}}(a - x) \\ \beta_1 &= -\sqrt{\frac{2}{\lambda z}}(a + y) \qquad & \beta_2 &= +\sqrt{\frac{2}{\lambda z}}(a - y) \end{aligned} \tag{3.43}$$

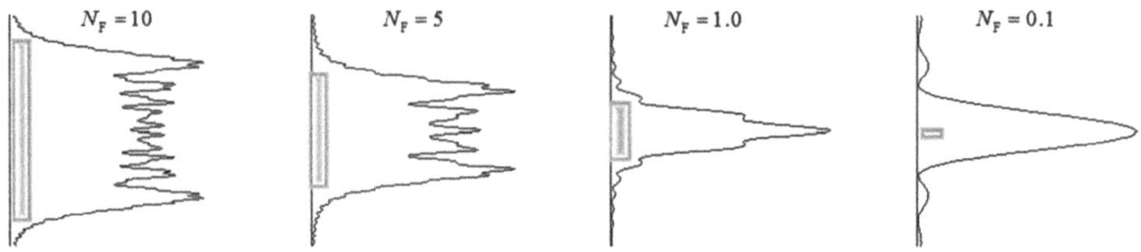

Figure 3.1. One-dimensional diffraction patterns for a uniform slit. The gray boxes represent the aperture size.

and the Fresnel integrals are

$$C(z) = \int_0^z \cos\left(\frac{\pi t^2}{2}\right)\mathrm{d}t \qquad S(z) = \int_0^z \sin\left(\frac{\pi t^2}{2}\right)\mathrm{d}t. \tag{3.44}$$

From the field, the intensity can be calculated.

$$I(x, y) = \frac{1}{4}\left\{\left[C(\alpha_2) - C(\alpha_1)\right]^2 + \left[S(\alpha_2) - S(\alpha_1)\right]^2\right\} \times\left\{\left[C(\beta_2) - C(\beta_1)\right]^2 + \left[S(\beta_2) - S(\beta_1)\right]^2\right\} \tag{3.45}$$

It is sometimes argued that an expression like this is *not a closed form*. However, the Fresnel sine and cosine integrals are tabulated functions and no more *open* than an expression that contains exp functions or basic trigonometric functions.

3.3.2 Fresnel number

An important parameter in this derivation, although not explicitly needed, is the concept of *Fresnel number*. It is defined simply as the ratio

$$N_F = \frac{a^2}{\lambda z} \tag{3.46}$$

which is a rough estimate of the distance away from the aperture considering the Fresnel approximation. Figure 3.1 shows a slice plot of the Fresnel diffraction patterns at various distances from the aperture, represented by Fresnel numbers. Notice how the number of ripples across the aperture is approximately equal to the Fresnel number until it is less than unity.

When z is small, as it is very near the aperture, the Fresnel number is very high. When the distance z is far from the aperture, the Fresnel number is small. When z is very large compared to the aperture size, the Fresnel number is very small and we can incorporate another approximation.

3.4 The Fraunhofer approximation

When the propagation distance is much larger than the physical extent of the aperture, that is, in the *far field*,

$$z \gg \frac{k}{2}\left(\xi^2 + \eta^2\right)_{\max}. \tag{3.47}$$

The quadratic phase factor in the Fresnel approximation, (3.30), reduces to unity.

$$\exp\left[\mathrm{j}\frac{k}{2z}(\xi^2+\eta^2)\right]\approx 1. \tag{3.48}$$

This has the effect of reducing the diffraction integral even further. The result, the Fraunhofer diffraction integral, shows that the far-field distribution, up to a multiplicative factor in front of the integral, is simply the two-dimensional Fourier transform of the field in the aperture

$$U(x, y)=\frac{\mathrm{e}^{\mathrm{j}kz}}{\mathrm{j}\lambda z}\exp\left[\mathrm{j}\frac{k}{2z}(x^2+y^2)\right]\int_{-\infty}^{\infty}\int\left\{U_{\mathrm{Aper}}(\xi,\eta)\right\} \times\exp\left[-\mathrm{j}\frac{2\pi}{\lambda z}(x\xi+y\eta)\right]\mathrm{d}\xi\,\mathrm{d}\eta \tag{3.49}$$

at spatial frequencies

$$f_x=\frac{x}{\lambda z}\qquad f_y=\frac{y}{\lambda z}. \tag{3.50}$$

The condition (3.47) can be relaxed somewhat if the wave passing through the aperture is a spherical wave converging toward the observation point, or if a positive lens produces that converging wave at or near the aperture.

3.4.1 A rectangular aperture

Because the Fourier transforms can be found for many different geometries and configurations of waves, the mathematics of diffraction can be simplified. For example, consider a wave going through a rectangular aperture represented by

$$t_{\mathrm{Aper}}(\xi,\eta)=\mathrm{rect}\left(\frac{\xi}{w_x}\right)\mathrm{rect}\left(\frac{\eta}{w_y}\right) \tag{3.51}$$

where the widths of the apertures in the x- and y-directions are w_x and w_y.

$$U(x, y)=\frac{\mathrm{e}^{\mathrm{j}kz}}{\mathrm{j}\lambda z}\exp\left[\mathrm{j}\frac{k}{2z}(x^2+y^2)\right]\int_{-w_x2^{-1}}^{w_x2^{-1}}U(\xi)\exp\left[-\mathrm{j}\frac{2\pi}{\lambda z}(x\xi)\right]\mathrm{d}\xi \times\int_{-w_y2^{-1}}^{w_y2^{-1}}U(\eta)\exp\left[-\mathrm{j}\frac{2\pi}{\lambda z}(y\eta)\right]\mathrm{d}\eta \tag{3.52}$$

which becomes

$$U(x, y)=\frac{\mathrm{e}^{\mathrm{j}kz}}{\mathrm{j}\lambda z}\exp\left[\mathrm{j}\frac{k}{2z}(x^2+y^2)\right]w_xw_y\,\mathrm{sinc}\left(\frac{w_xx}{\lambda z}\right)\mathrm{sinc}\left(\frac{w_yy}{\lambda z}\right). \tag{3.53}$$

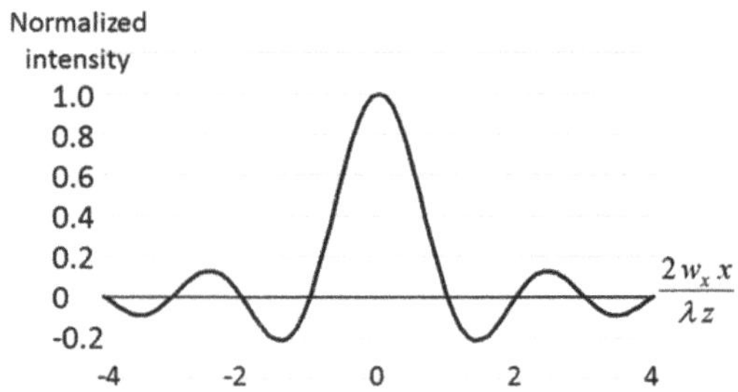

Figure 3.2. Cross-section of the Fraunhofer diffraction pattern of a rectangular aperture.

The intensity distribution at z is

$$I(x, y) = \left(\frac{w_x w_y}{\lambda z}\right)^2 \text{sinc}^2\left(\frac{w_x x}{\lambda z}\right)\text{sinc}^2\left(\frac{w_y y}{\lambda z}\right). \tag{3.54}$$

A cross-sectional plot of this distribution at $y = 0$ is given in figure 3.2 where the width of the central lobe is

$$\Delta x_{\text{Lobe}} = \frac{2\lambda z}{w_x}. \tag{3.55}$$

Note that the zeroes of the sinc functions of (3.54) will be equally spaced from the origin, thus the positions of the maxima and minima are repeated. Note also that the width of the central lobe, and actually all of the distribution, is inversely proportional to the width of the aperture. Large apertures produce small diffraction patterns and vice versa.

3.4.2 Circular aperture

For a circular aperture, we can make use of the Fourier–Bessel transforms to simplify the diffraction integral. For a circular aperture of radius R with q being the polar coordinate in the plane of the aperture,

$$t_{\text{Aper}}(q) = \text{circ}\left(\frac{q}{R}\right). \tag{3.56}$$

The field in the observation plane is

$$U(r) = \frac{e^{jkz}}{j\lambda z} \exp\left(j\frac{kr^2}{2z}\right) \text{FB}[U(q)]|_{\rho=r/\lambda z}. \tag{3.57}$$

When we use the Fourier–Bessel identity

$$\text{FB}\left[\text{circ}\left(\frac{q}{R}\right)\right] = \pi R^2 \frac{J_1(2\pi R\rho)}{\pi R\rho} \tag{3.58}$$

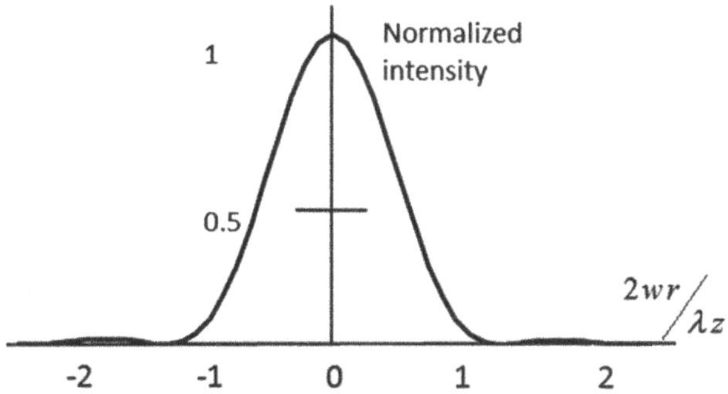

Figure 3.3. Cross-section of the Fraunhofer diffraction pattern of a circular aperture.

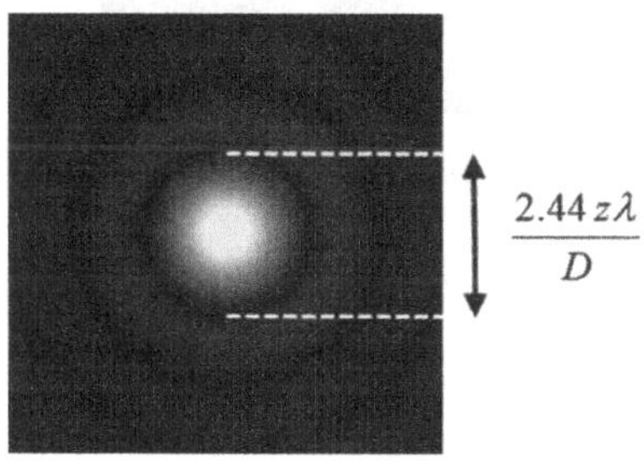

Figure 3.4. Fraunhofer diffraction pattern of a circular aperture of diameter D (Airy pattern). The central bright spot is the Airy disk.

the result is the far-field distribution.

$$U(r) = \frac{\pi R^2 \mathrm{e}^{\mathrm{j}kz}}{\mathrm{j}\lambda z} \exp\left(\mathrm{j}\frac{kr^2}{2z}\right) \frac{2J_1(kRr/z)}{kRr/z}. \tag{3.59}$$

The corresponding intensity distribution is

$$I(r) = \left(\frac{\pi R^2}{\lambda z}\right)^2 \left[\frac{2J_1(kRr/z)}{kRr/z}\right]^2. \tag{3.60}$$

The distribution is called the Airy pattern with a central disk (between the first zeroes of the function) equal to

$$d_{\mathrm{Airy}} = 2.44\frac{\lambda z}{R}. \tag{3.61}$$

Figures 3.3 and 3.4 show the patterns and the circular symmetry that extends from the aperture to the far field. Note that the zeroes of the Bessel function are not equally spaced like the sinc functions of (3.54), so the positions of the extrema rings are concentric but not equally spaced.

With the help of computers, the Fourier transforms of countless aperture geometries can be calculated. The fields in (3.49) can be computed and the

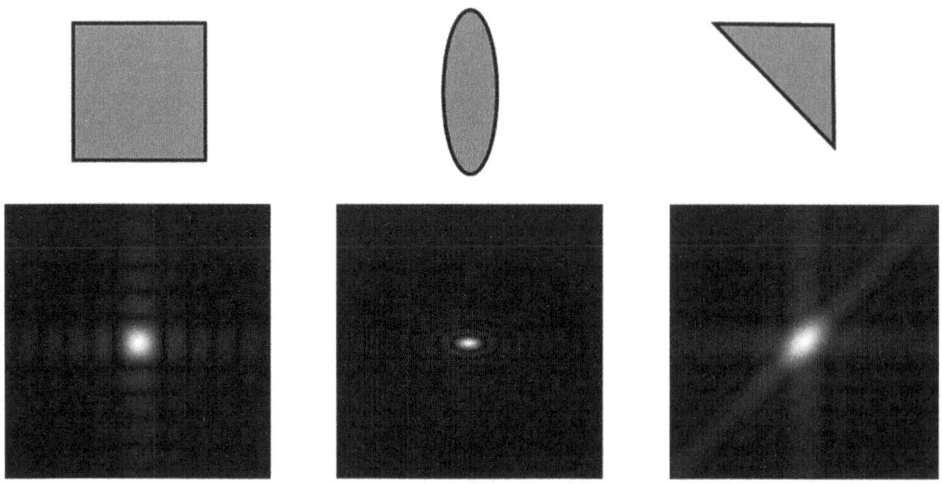

Figure 3.5. Aperture shapes and their Fraunhofer diffraction patterns. (a) Square, (b) ellipse and (c) triangle.

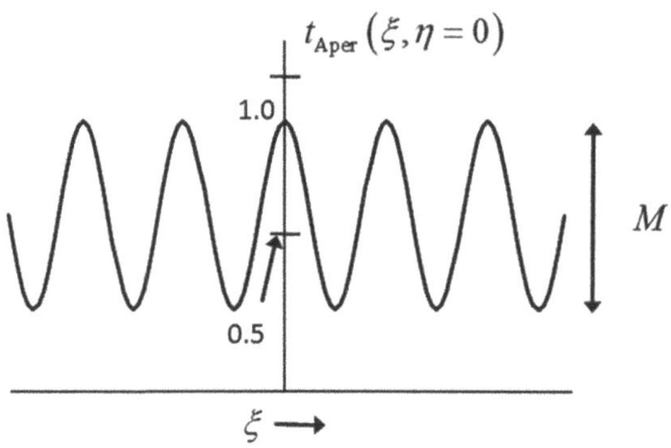

Figure 3.6. The amplitude transmission function of a sinusoidal amplitude grating with modulation M.

corresponding intensity patterns can be displayed. Figure 3.5 shows various aperture shapes and their corresponding far-field intensity distributions.

3.4.3 Thin sinusoidal amplitude grating

A number of Fraunhofer diffraction patterns can be computed in a closed form that allows us to identify and characterize various aspects of the diffraction process. One such case is where the field across the aperture is a slowly varying sinusoidal amplitude grating. The transmission function through a square aperture of side w is given by

$$t_{\mathrm{Aper}}(\xi, \eta) = \left[\frac{1}{2} + \frac{M}{2}\cos(2\pi f_0 \xi)\right]\mathrm{rect}\left(\frac{\xi}{w}\right)\mathrm{rect}\left(\frac{\eta}{w}\right). \tag{3.62}$$

This is plotted in figure 3.6 where the modulation of the cosine wave is M and f_0 is the spatial frequency of the modulation along the ξ-direction. Invoking the

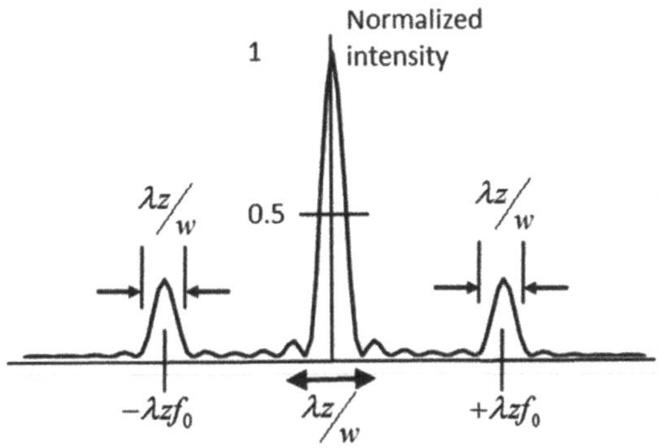

Figure 3.7. Fraunhofer diffraction pattern for a thin sinusoidal amplitude grating.

separability of the problem, the various Fourier transform pairs for the rect( ) function and the convolution theorem for the product of Fourier transforms, the resultant diffraction pattern intensity becomes

$$I(x, y) \approx \left[\frac{w^2}{2\lambda z}\right]^2 \mathrm{sinc}^2\left(\frac{wy}{\lambda z}\right)\left\{\mathrm{sinc}^2\left(\frac{wx}{\lambda z}\right) + \frac{M^2}{4}\mathrm{sinc}^2\left[\frac{w}{\lambda z}(x + f_0\lambda z)\right] + \frac{M^2}{4}\mathrm{sinc}^2\left[\frac{w}{\lambda z}(x - f_0\lambda z)\right]\right\}. \tag{3.63}$$

The 'approximately equal' sign is invoked because we are assuming that the modulation is slowly varying, that is $f_0 \gg \frac{2}{w}$, and the three sinc functions do not overlap. A cross-sectional plot of (3.63) is shown is figure 3.7.

3.4.4 Thin sinusoidal phase grating

A final closed-form solution that we will consider is a thin sinusoidal *phase* grating. The distribution directly after the aperture is given by

$$t_{\mathrm{Aper}}(\xi, \eta) = \exp\left[\mathrm{j}\frac{M}{2}\sin(2\pi f_0\xi)\right]\mathrm{rect}\left(\frac{\xi}{w}\right)\mathrm{rect}\left(\frac{\eta}{w}\right). \tag{3.64}$$

If we again assume $f_0 \gg \frac{2}{w}$, the intensity pattern can be calculated.

$$I(x, y) \approx \left[\frac{w^2}{\lambda z}\right]^2 \sum_{q=-\infty}^{\infty} J_q^2\left(\frac{M}{2}\right)\mathrm{sinc}^2\left(\frac{wy}{\lambda z}\right)\mathrm{sinc}^2\left[\frac{w}{\lambda z}(x - qf_0\lambda z)\right]. \tag{3.65}$$

In this phase grating case, the peak intensity of the qth order is $[\frac{w^2 J_q(M2^{-1})}{\lambda z}]^2$ and its distance from the origin (center of the diffraction pattern) is $qf_0\lambda z$. Another parameter of note is the diffraction efficiency η_q, that is, how much power is diffracted into a particular order q.

$$\eta_q = J_q^2\left(\frac{M}{2}\right) \tag{3.66}$$

References

Goodman J W 2005 *Introduction to Fourier Optics* 3rd edn (Greenwood Village, CO: Roberts)

Sherman G C 1967 Application of the convolution theorem to Rayleigh's integral formulas *J. Opt. Soc. Am.* **57** 546–7

Chapter 4

Fourier transforms and optics

4.1 Fourier transforming properties of lenses

Other than a plane mirror, the simplest and most ancient optical element is a lens. A piece of glass or plastic or amber or anything transparent can be utilized as a lens. Its purpose is to focus light which essentially moves an image at infinity to an image at some finite distance. A thin lens is an approximation whereby a field impinging upon a lens, with the z-axis normal to the front surface of the lens, does not have a noticeable shift of the x- or y-coordinates. With the material in the lens having an index of refraction n, the phase delay is given by

$$\phi(x, y) = \frac{2\pi}{\lambda} n\Delta(x, y) + \frac{2\pi}{\lambda}\left[\Delta_0 - \Delta(x, y)\right]. \tag{4.1}$$

Because we want to represent the lens in a wave-optics sense, the lens may also be a simple multiplicative phase change of the field

$$t_{\mathrm{Lens}} = \exp\left[jk\Delta_0\right]\exp[jk(n-1)\Delta(x, y)] \tag{4.2}$$

where $k = \frac{2\pi}{\lambda}$ and $\Delta(x, y)$ is the shape of the lens, often called the thickness function of the lens. See figure 4.1.

The field after the lens U'_{Lens} is simply

$$U'_{\mathrm{Lens}}(x, y) = t_{\mathrm{Lens}}(x, y)\,U_{\mathrm{Lens}}(x, y). \tag{4.3}$$

In the paraxial approximation where spherical surfaces are represented by parabolic surfaces,

$$\Delta(x, y) = \Delta_0 - \frac{x^2 + y^2}{2}\left(\frac{1}{R_1} - \frac{1}{R_2}\right). \tag{4.4}$$

doi:10.1088/978-0-750-31056-7ch4

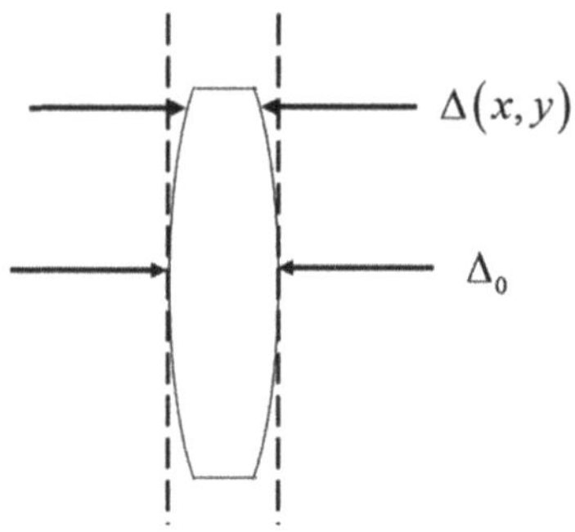

Figure 4.1. The thickness function for a lens.

Table 4.1. The six geometric configurations of lens surfaces.

Type	R_1	R_2	f
Double convex	>0	<0	>0
Plano-convex	>0	∞	>0
Positive meniscus	>0	$>0, \lvert R_2\rvert > \lvert R_1\rvert$	>0
Double concave	<0	>0	<0
Plano-concave	<0	∞	<0
Negative meniscus	<0	$>0, \lvert R_2\rvert < \lvert R_1\rvert$	<0

Considering figure 4.1, for rays traveling from left to right, a convex surface, such as that shown, has a positive radius of curvature ($R_1 > 0$), and the second (rightmost) surface is concave and has a negative radius of curvature ($R_2 < 0$).

Substituting (4.4) into (4.2) gives

$$t_{\text{Lens}} = \exp\left[\mathrm{j}kn\Delta_0\right]\exp\left[-\mathrm{j}k(n-1)\frac{x^2+y^2}{2}\left(\frac{1}{R_1}-\frac{1}{R_2}\right)\right] \tag{4.5}$$

which leads to a useful definition of the focal length f of a lens:

$$\frac{1}{f} \equiv (n-1)\left(\frac{1}{R_1}-\frac{1}{R_2}\right). \tag{4.6}$$

The constant factor $\exp[\mathrm{j}kn\Delta_0]$ can usually be neglected such that

$$t_{\text{Lens}} = \exp\left[-\mathrm{j}\frac{k}{2f}\frac{x^2+y^2}{2}\right] \tag{4.7}$$

can represent a thin lens. The finite extent of the lens $x_{\max}$ and $y_{\max}$ is considered later and the six geometric configurations of surfaces can be found in table 4.1.

A thin lens has the effect of changing the vergence (*con*vergence or *di*vergence) of a wave going through it. For an incoming plane wave (vergence is zero), $U_{\text{Lens}}(x, y) = 1$, and the field exiting the lens is

$$U'_{\text{Lens}}(x, y) = t_{\text{Lens}} = \exp\left[-\mathrm{j}\frac{k}{2f}\frac{x^2 + y^2}{2}\right]. \tag{4.8}$$

A positive lens ($f > 0$) would lead to a convergent or focused beam in the positive z region and a negative lens ($f < 0$) would lead to a convergent or focused beam in the negative z region.

4.2 Coherence and Fourier transforming

Another property of lenses is their apparent ability to take two-dimensional Fourier transforms at literally the speed of light. We first consider monochromatic, coherent systems where the fields are linear in complex amplitude. The input is a two-dimensional function of light which could be from either a transmitted or reflected source. This input field is called the *object*. We want to find out what a lens does and how it produces an output. Three configurations are of interest: (i) the object is placed directly at the lens, (ii) the object is placed to the left (in front) of the lens and (iii) the object is placed to the right of (behind) the lens. See figure 4.2.

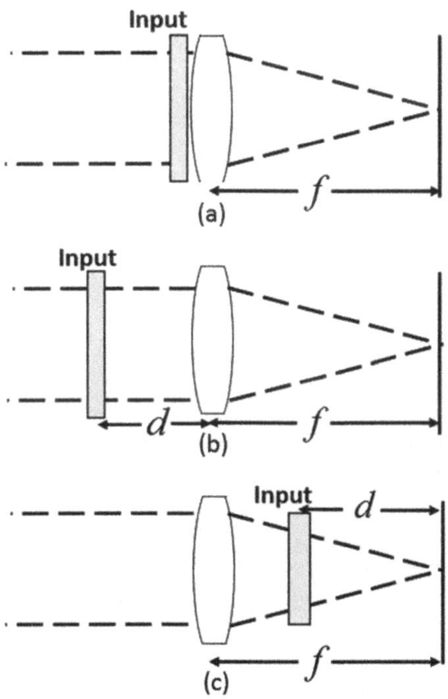

Figure 4.2. Geometries for performing a Fourier transform with a positive lens. (*a*) the input is placed at the lens, (*b*) the input is before the lens, and (*c*) the input is after the lens.

4.2.1 Input placed against the lens

The input is illuminated by a normally incident, monochromatic plane wave of amplitude A. The field at the lens in figure 4.2(a) is

$$U_{\text{Lens}}(x, y) = A\, t_{\text{Lens}}(x, y). \tag{4.9}$$

The light to the lens is restricted by a pupil, where the pupil function is equal to unity inside the lens (pupil) and zero elsewhere.

$$\mathrm{P}(x, y) = \begin{cases} 1 & \text{inside the pupil} \\ 0 & \text{otherwise.} \end{cases} \tag{4.10}$$

The amplitude directly behind the lens is

$$U'_{\text{Lens}}(x, y) = \mathrm{P}(x, y)\, U_{\text{Lens}}(x, y) \exp\left[-\mathrm{j}\frac{k}{2f}\frac{x^2 + y^2}{2}\right]. \tag{4.11}$$

With Cartesian coordinates (u, v) we use the Fresnel diffraction integral from chapter 3 with (4.11) substituted. The result is

$$U(u, v) = \frac{\mathrm{e}^{\mathrm{j}kz}}{\mathrm{j}\lambda z} \exp\left[\mathrm{j}\frac{k}{2z}(u^2 + v^2)\right] \int_{-\infty}^{\infty}\!\!\int \mathrm{P}(x, y)\, U_{\text{Lens}}(x, y)$$
$$\times \exp\left[-\mathrm{j}\frac{k}{2f}\frac{x^2 + y^2}{2}\right] \exp\left[-\mathrm{j}\frac{2\pi}{\lambda z}(xu + yv)\right] \mathrm{d}x\, \mathrm{d}y. \tag{4.12}$$

To find the field at the back focal plane ($z = f$), we drop the constant phase factor and we see that the quadratic exponential terms exactly cancel. The resultant field at $z = f$ is

$$U_f(u, v) = \frac{1}{\mathrm{j}\lambda f} \exp\left[\mathrm{j}\frac{k}{2f}(u^2 + v^2)\right] \int_{-\infty}^{\infty}\!\!\int \mathrm{P}(x, y)\, U_{\text{Lens}}(x, y)$$
$$\times \exp\left[-\mathrm{j}\frac{2\pi}{\lambda f}(xu + yv)\right] \mathrm{d}x\, \mathrm{d}y. \tag{4.13}$$

In many cases, the input is physically smaller than the lens, or at least we are only considering the light entering the lens and not missing it. In that case $\mathrm{P}(x, y) \to 1$ and

$$U_f(u, v) = \frac{\mathrm{e}^{\mathrm{j}\frac{k}{2f}(u^2+v^2)}}{\mathrm{j}\lambda f} \int_{-\infty}^{\infty}\!\!\int U_{\text{Lens}}(x, y) \exp\left[-\mathrm{j}\frac{2\pi}{\lambda f}(xu + yv)\right] \mathrm{d}x\, \mathrm{d}y. \tag{4.14}$$

We see that the complex field at the focal plane of the lens is simply the Fourier transform of the field at the lens (at spatial frequencies $f_x = u/\lambda f, f_y = v/\lambda f$) multiplied by a quadratic phase factor. It is also apparent that the field at the focal plane is also the Fraunhofer diffraction pattern of the field at the lens even though we are at a normally small distance f from the lens and nowhere near the distance required by the Fraunhofer approximation.

If we do not care about the phase factors as they propagate to the focal plane and possibly to other planes as well, we can examine the intensity distribution at the focal plane which is equivalent to the *energy spectrum* of the input.

$$I_f(u, v) = \left(\frac{A}{\lambda f}\right)^2 \left| \int_{-\infty}^{\infty}\int U_{\text{Lens}}(x, y) \exp\left[-\text{j}\frac{2\pi}{\lambda f}(xu + yv)\right] \text{d}x\, \text{d}y \right|^2. \quad (4.15)$$

4.2.2 Input placed in front of the lens

Now consider the case where the input transparency, the object, is at a distance d in front of the lens. See figure 4.2(b). We can make use of the Fourier aspects of the Fresnel approximation that relate the spectrum of field at the lens $\text{FT}_{\text{Lens}}(f_x, f_y)$ after propagation distance d from the object

$$\text{FT}_{\text{Lens}}(f_x, f_y) = \text{FT}_{\text{Obj}}(f_x, f_y) \exp\left[-\text{j}\pi\lambda d\left(f_x^2 + f_y^2\right)\right]. \quad (4.16)$$

The object is in the ξ–η plane; the lens is in the x–y plane; and the output is in the u–v plane.

$$\text{FT}_{\text{Lens}}(f_x, f_y) = \text{FT}\left[U_{\text{Lens}}(x, y)\right] \quad (4.17)$$

and

$$\text{FT}_{\text{Obj}}(f_x, f_y) = \text{FT}\left[A\, t_{\text{Obj}}(\xi, \eta)\right]. \quad (4.18)$$

Propagating to the output plane, the field is

$$U_f(u, v) = \frac{\exp\left[\text{j}\frac{\pi}{\lambda f}(u^2 + v^2)\right]}{\text{j}\lambda f} \text{FT}_{\text{Lens}}\left(\frac{u}{\lambda f}, \frac{v}{\lambda f}\right). \quad (4.19)$$

Substituting (4.16)–(4.18) into (4.19), we have the result

$$U_f(u, v) = \frac{A \exp\left[\text{j}\frac{\pi}{\lambda f}\left(1 - \frac{d}{f}\right)(u^2 + v^2)\right]}{\text{j}\lambda f} \int\int_{-\infty}^{\infty} t_{\text{Obj}}(\xi, \eta) \times \exp\left[-\text{j}\frac{2\pi}{\lambda f}(\xi u + \eta v)\right] \text{d}\xi\, \text{d}\eta. \quad (4.20)$$

The phase factor that precedes the integral has a very important property: it vanishes when $d = f$. That is, when the object is placed at the front focal plane of the lens, phase curvature disappears, and the field in the back focal plane of the lens is exactly the Fourier transform of the object field.

It is possible that some of the light from the object does not pass through the lens because of its finite extent. This limitation, known as vignetting, can distort the Fourier transform that is observed. This can be accommodated in (4.20) by introducing a pupil function $P(\xi, \eta)$ centered at coordinates $\xi = -\frac{du}{f}, \eta = -\frac{dv}{f}$.

$$U_f(u, v) = \frac{A \exp\left[j\frac{\pi}{\lambda f}\left(1 - \frac{d}{f}\right)(u^2 + v^2)\right]}{j\lambda f} \int\int_{-\infty}^{\infty} t_{\text{Obj}}(\xi, \eta)\, P\left(\xi + \frac{d}{f}u, \eta + \frac{d}{f}v\right) \times \exp\left[-j\frac{2\pi}{\lambda f}(\xi u + \eta v)\right] d\xi\, d\eta \tag{4.21}$$

4.2.3 Input placed behind the lens

If the object is placed a distance *d* behind the lens, it is illuminated with a converging beam with a size reduced by the factor (figure 4.2(*c*)). Again using the Fresnel diffraction integral, the field at the back focal plane is

$$U_f(u, v) = \frac{A \exp\left[j\frac{\pi}{\lambda d}(u^2 + v^2)\right]}{j\lambda d} \frac{f}{d} \int\int_{-\infty}^{\infty} t_{\text{Obj}}(\xi, \eta)\, P\left(\xi\frac{f}{d}, \eta\frac{f}{d}\right) \times \exp\left[-j\frac{2\pi}{\lambda d}(\xi u + \eta v)\right] d\xi\, d\eta. \tag{4.22}$$

4.3 Monochromatic image formation

We now know that a lens can form a Fourier transform of an object distribution. We will now show how a lens can form a replica of the object, called an *image*. First we consider only monochromatic illumination and an aberration-free lens. To begin simply, let us look at how a lens deals with the simplest of objects, a point of light.

4.3.1 The impulse response of a positive lens

Going beyond the condition where we look at the output field at the focal plane of the lens, we will look at the output at an arbitrary distance behind the lens. The object field $U_{\text{Obj}}(\xi, \eta)$ is a distance z_1 in front of a positive lens. We want to find the field $U_{\text{img}}(u, v)$ at a distance z_2 behind the lens that represents a field similar to the object, with only a linear magnification of coordinates, which we call an *image*. See figure 4.3. Assuming linearity, we will find the superposition integral

$$U_{\text{img}}(u, v) = \int_{-\infty}^{\infty}\int h(u, v; \xi, \eta)\, U_{\text{Obj}}(\xi, \eta)\, d\xi\, d\eta. \tag{4.23}$$

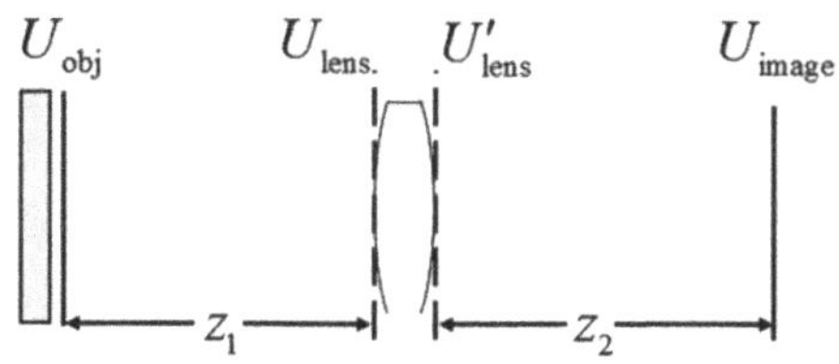

Figure 4.3. Geometry for image formation.

Here, $h(u, v; \xi, \eta)$ is the impulse response of the system, that is, the field at (u, v) from a point source at $\xi = \eta = 0$. To find $h(;)$ we use a delta function $\delta(\xi, \eta)$ to represent the point source. The field at the lens is a spherical wave diverging from the point.

$$U_{\text{Lens}}(x, y) = \frac{1}{j\lambda z_1} \exp\left\{ j\frac{\pi}{\lambda z_1}\left[(x-\xi)^2 + (y-\eta)^2\right]\right\}. \tag{4.24}$$

After passage through the lens of focal length f, with pupil function $P(x, y)$ describing the physical extent and nature of the lens, the field on the output side of the lens is

$$U'_{\text{Lens}}(x, y) = U_{\text{Lens}}(x, y)P(x, y)\exp\left[-j\frac{\pi}{\lambda f}(x^2 + y^2)\right]. \tag{4.25}$$

Using the Fresnel approximation to propagate the field to the z_2 plane, we have a complicated result with quadratic phase factors and Fourier transforms.

$$\begin{aligned} h(u, v; \xi, \eta) = {} & \frac{1}{\lambda^2 z_1 z_2}\exp\left[j\frac{\pi}{\lambda z_2}(u^2 + v^2)\right]\exp\left[j\frac{\pi}{\lambda z_1}(\xi^2 + \eta^2)\right] \\ & \times \int_{-\infty}^{\infty}\int P(x, y)\frac{1}{\lambda^2 z_1 z_2}\exp\left[j\frac{\pi}{\lambda}\left(\frac{1}{z_1} + \frac{1}{z_2} - \frac{1}{f}\right)(x^2 + y^2)\right] \\ & \times \exp\left\{-j\frac{2\pi}{\lambda}\left[\left(\frac{\xi}{z_1} + \frac{u}{z_2}\right)x + \left(\frac{\eta}{z_1} + \frac{v}{z_2}\right)y\right]\right\}dx\,dy \end{aligned} \tag{4.26}$$

The quadratic phase factors at the front of the equation are independent of lens coordinates (x, y) or lens property f. The quadratic phase factor inside the integral can be eliminated if we set $\frac{1}{z_1} + \frac{1}{z_2} - \frac{1}{f} = 0$, the classical geometrical optics lens law. Also, we can restrict the imaging process to just considering the intensity at plane z_2 or keeping the phase term but evaluating the image on a spherical surface of radius z_2 centered on the lens surface at the optical axis. If we further define the imaging system magnification to be

$$M = -\frac{z_2}{z_1}, \tag{4.27}$$

we have a simplified result for the impulse response. The negative sign in (4.27) indicates that the image from a single positive lens is inverted, as we know it to be. Finally,

$$h(u,\, v;\, \xi,\, \eta) \approx \frac{1}{\lambda^2 z_1 z_2} \int_{-\infty}^{\infty} \int \mathrm{P}(x,\, y) \exp\left\{ -\mathrm{j}\frac{2\pi}{\lambda z_2}[(u - M\xi)x + (v - M\eta)y] \right\} \mathrm{d}x\, \mathrm{d}y \tag{4.28}$$

which is seen as the Fraunhofer diffraction pattern of the lens aperture centered at $u = M\xi$, $v = M\eta$. We rightly conclude that the response of an optical system to a point source, where the lens law is satisfied, is just the Fourier transform of the aperture.

By expanding this solution, we can find out how a full image field is produced. We assume that the object, or the light from it, is made up of multiple point sources, each with a different amplitude. That is, the object is a product of the light distribution, the geometrical object $U_{\text{geom}}(u,\, v)$ and a Dirac delta function $\delta(u,\, v)$. If we assume space-invariance, our image will be a magnified, or minified, version of the object.

$$U_{\text{geom}}(u,\, v) = \frac{1}{|M|} U_{\text{Obj}}\left(\frac{u}{M},\, \frac{v}{M}\right) \tag{4.29}$$

Making a set of normalizing coordinate transformations

$$\begin{aligned} \tilde{\xi} &= M\xi \\ \tilde{\eta} &= M\eta \\ \tilde{x} &= \frac{x}{\lambda z_2} \\ \tilde{y} &= \frac{y}{\lambda z_2} \\ \tilde{h} &= \frac{h}{|M|} \end{aligned} \tag{4.30}$$

the impulse response becomes

$$h\left(u,\, v;\, \tilde{\xi},\, \tilde{\eta}\right) = \frac{1}{\lambda^2 z_1 z_2} \int_{-\infty}^{\infty} \int \mathrm{P}(x,\, y) \exp\left\{ -\mathrm{j}\frac{2\pi}{\lambda z_2}\left[(u - \tilde{\xi})x + (v - \tilde{\eta})y\right] \right\} \mathrm{d}x\, \mathrm{d}y \tag{4.31}$$

and the image field becomes

$$U_{\text{img}}(u,\, v) = \int_{-\infty}^{\infty} \int h\left(u - \tilde{\xi},\, v - \tilde{\eta}\right) \left[\frac{1}{|M|} U_{\text{Obj}}\left(\frac{\xi}{M},\, \frac{\tilde{\eta}}{M}\right) \right] \mathrm{d}\tilde{\xi}\, \mathrm{d}\tilde{\eta}. \tag{4.32}$$

Equation (4.32) can be recognized as a convolution of the geometrical object (4.29) and the normalized impulse response $\tilde{h}(u,\, v)$, also known as the PSF of the optical system, which happens to be the Fraunhofer diffraction pattern of the lens and pupil. The PSF is given by

$$\tilde{h}(u, v) = \int_{-\infty}^{\infty}\int P(\lambda z_2\tilde{x}, \lambda z_2\tilde{y}) \exp\left[-j2\pi(u\tilde{x} + v\tilde{y})\right] d\tilde{x}\, d\tilde{y} \tag{4.33}$$

and the entire image formation operation can be summarized

$$U_{\text{img}}(u, v) = \tilde{h}(u, v) \otimes U_{\text{geom}}(u, v). \tag{4.34}$$

We physically experience this convolution operation every time we observe an image. The detailed information in the object is smoothed by the convolution operation and we lose some of that information, which cannot be recovered. The detail, represented by the modulation of light above or below the average illumination, is reduced. Looking back over this chapter, we can see how a large PSF, from primarily a small aperture, will wash out the object detail by virtue of (4.34). We will discuss that in detail in the next chapter.

IOP Publishing

Principles and Applications of Fourier Optics

Robert K Tyson

Chapter 5

Imaging systems and aberrations

We can extend the discussion of imaging from a single lens to imaging from a complex system of lenses, mirrors and many other optical elements. By collecting all the optics into a 'system' we can describe the 'system' with an entrance pupil (where the light goes in from the object a distance z_{Obj} in front of it), an exit pupil (where the light comes out and travels a distance z_{img} to the image) and a focal length f.

The assumptions of the previous chapter include monochromatic sources of light. Light is almost *never* ideally monochromatic; even lasers have a certain finite bandwidth. The inclusion of varying time can be used to examine the effects of polychromatic light as was shown in chapter 3.

From equation (4.34), we see that an image is formed by adding together, via the convolution operation, the point sources from points of light in the object. While each source point may differ in relative phase from another point, the absolute phase may be correlated from point to point. This is called a *spatially coherent* (or simply 'coherent') source. When the phases from all points in the object are not correlated in any way, the source is said to be *spatially incoherent* ('incoherent').

When the light from the object is coherent, the complex amplitude impulse responses are added to form the image. *A coherent imaging system is linear in complex amplitude.*

With incoherent light, the impulse response is the squared magnitude of the amplitude impulse response (its intensity). Because the phases are varying without any correlation, the image is made up of a sum of intensities from the point sources. *An incoherent imaging system is linear in intensity.*

For narrowband conditions, the time-dependent image field is

$$U_{\mathrm{img}}(u,\, v;\, t) = \int_{-\infty}^{\infty}\int h\left(u - \tilde{\xi},\, v - \tilde{\eta}\right) U_{\mathrm{geom}}\left(\tilde{\xi},\, \tilde{\eta};\, t - \tau\right) \mathrm{d}\tilde{\xi}\; \mathrm{d}\tilde{\eta} \qquad (5.1)$$

doi:10.1088/978-0-750-31056-7ch5

where τ is the time delay between the point $(\tilde{\xi}, \tilde{\eta})$ and the point (u, v). The coherent image intensity is the time average of the instantaneous intensity $U_{img}(u, v; t)$.

$$I_{img}(u, v) = \left\langle \left| U_{img}(u, v; t) \right|^2 \right\rangle = \left| \int_{-\infty}^{\infty} \int h(u - \tilde{\xi}, v - \tilde{\eta}) U_{geom}(\tilde{\xi}, \tilde{\eta}) \, d\tilde{\xi} \, d\tilde{\eta} \right|^2. \quad (5.2)$$

When the object is perfectly incoherent, the image intensity is a convolution of the intensity impulse response $|h|^2$ with the ideal geometric image, I_{geom}, or

$$I_{img}(u, v) = \kappa \int_{-\infty}^{\infty} \int \left| h(u - \tilde{\xi}, v - \tilde{\eta}) \right|^2 I_{geom}(\tilde{\xi}, \tilde{\eta}) \, d\tilde{\xi} \, d\tilde{\eta}. \quad (5.3)$$

The constant κ is inserted to cover various real light intensity levels. When the source is neither completely coherent nor completely incoherent, it is said to be partially coherent and the Fourier approach becomes considerably more complex. If however, there is an extended incoherent source, such as the light from the surface of the Sun, it can behave like an incoherent source or a coherent source, dependent upon the pupil's distance from the source. We define angles from the object plane where θ_{pupil} is the angular diameter of the entrance pupil of the imaging system, θ_{source} is the angular diameter of the source and θ_{object} is the angular diameter of the angular spectrum of the object. If $\theta_{source} \geqslant \theta_{object} + \theta_{pupil}$, the imaging system will behave like an incoherent system, but if $\theta_{source} \ll \theta_{pupil}$, the system will behave like an coherent system. We experience this phenomenon daily. The Sun, or planets in our solar system, appear to be extended sources and are incoherent objects. Stars, which are essentially the same as the Sun but much further away, appear as coherent objects. Atmospheric twinkling of stars is only visible for coherent objects, such that stars and planets are easily distinguishable from one another.

5.1 Coherent image formation and the amplitude transfer function

We can now take advantage of the Fourier aspects of the convolution operation. Repeating the equation for coherent image formation, we see that the image field is the convolution of the amplitude PSF and the object that would result solely from geometric optics considerations.

$$U_{img}(u, v) = \tilde{h}(u, v) \otimes U_{geom}(u, v) \quad (5.4)$$

By Fourier transforming both sides of this equation, we can find that the frequency spectrum of the image field is the product of the spectrum of the impulse response times the spectrum of the object.

$$\mathrm{FT}(U_{img}) = \mathrm{FT}(h)\mathrm{FT}(U_{geom}) \quad (5.5)$$

$$\mathrm{FT}(U_{geom}) = \int\int_{-\infty}^{\infty} U_{geom}(u, v) \exp[-j2\pi(f_x u + f_y v)] du \, dv \quad (5.6)$$

$$\mathrm{FT}(U_{\mathrm{img}}) = \iint_{-\infty}^{\infty} U_{\mathrm{img}}(u, v) \exp[-\mathrm{j}2\pi(f_x u + f_y v)]\mathrm{d}u\,\mathrm{d}v \tag{5.7}$$

FT(h) has a special name. It is called the amplitude transfer function and will be denoted as $H_{\mathrm{coh}}(f_x, f_y)$.

$$H_{\mathrm{coh}}(f_x, f_y) = \iint_{-\infty}^{\infty} h(u, v) \exp[-\mathrm{j}2\pi(f_x u + f_y v)]\mathrm{d}u\,\mathrm{d}v \tag{5.8}$$

We recognize $H_{\mathrm{coh}}(f_x, f_y)$ as the Fourier transform of the pupil function.

$$H_{\mathrm{coh}}(f_x, f_y) = \mathrm{FT}\left\{\frac{A}{\lambda z_{\mathrm{img}}}\iint_{-\infty}^{\infty} \mathrm{P}(x, y) \exp\left[-\mathrm{j}\frac{2\pi}{\lambda z_{\mathrm{img}}}(ux + vy)\right]\mathrm{d}x\,\mathrm{d}y\right\}. \tag{5.9}$$

A Fourier transform *of a Fourier transform of a function* is a scaled version of the function itself. Thus the amplitude transfer function can be written as a scaled pupil function.

$$H_{\mathrm{coh}}(f_x, f_y) = (A\lambda z_{\mathrm{img}})\mathrm{P}(-\lambda z_{\mathrm{img}} f_x, -\lambda z_{\mathrm{img}} f_y) \tag{5.10}$$

In most cases, the coordinates x and y are symmetrical, so we can eliminate the negative sign and also scale the constant $A\lambda z_{\mathrm{img}}$ to unity, thus

$$H_{\mathrm{coh}}(f_x, f_y) = \mathrm{P}(\lambda z_{\mathrm{img}} f_x, \lambda z_{\mathrm{img}} f_y). \tag{5.11}$$

This has a very interesting property. For coherent imaging, if the pupil is like many truly physical pupils, it has a value of one inside it and a value of zero outside it. Therefore the pupil cleanly defines which spatial frequencies are passed from the object to the image and which are not.

For a circular pupil of radius R, $\mathrm{P}(x, y) = \mathrm{P}(r) = \mathrm{circ}(\frac{r}{2R})$ and the amplitude transfer function is

$$H_{\mathrm{coh}}(f_x, f_y) = \mathrm{circ}\left(\frac{\sqrt{f_x^2 + f_y^2}}{R(\lambda z_{\mathrm{img}})^{-1}}\right). \tag{5.12}$$

The cutoff frequency f_{cutoff} is the largest spatial frequency passed along to the image. It is where the circ( ) function goes to zero, or

$$f_{\mathrm{cutoff}} = \frac{R}{\lambda z_{\mathrm{img}}}. \tag{5.13}$$

5.2 Incoherent image formation and the optical transfer function

Incoherent imaging systems follow the intensity convolution integral (5.3)

$$I_{\mathrm{img}}(u, v) = \kappa\int_{-\infty}^{\infty}\int \left|h(u - \tilde{\xi}, v - \tilde{\eta})\right|^2 I_{\mathrm{geom}}(\tilde{\xi}, \tilde{\eta})\,\mathrm{d}\tilde{\xi}\,\mathrm{d}\tilde{\eta}. \tag{5.14}$$

To remove the constant, we normalize the spectra by the image intensity levels.

$$\mathrm{FT}[I_{\mathrm{geom}}(f_x, f_y)] = \frac{\iint_{-\infty}^{\infty} I_{\mathrm{geom}}(u, v)\exp[-\mathrm{j}2\pi(f_x u + f_y v)]\mathrm{d}u\,\mathrm{d}v}{\iint_{-\infty}^{\infty} I_{\mathrm{geom}}(u, v)\mathrm{d}u\,\mathrm{d}v} \tag{5.15}$$

$$\mathrm{FT}[I_{\mathrm{img}}(f_x, f_y)] = \frac{\iint_{-\infty}^{\infty} I_{\mathrm{img}}(u, v)\exp[-\mathrm{j}2\pi(f_x u + f_y v)]\mathrm{d}u\,\mathrm{d}v}{\iint_{-\infty}^{\infty} I_{\mathrm{img}}(u, v)\mathrm{d}u\,\mathrm{d}v} \tag{5.16}$$

These spectra have a maximum at the origin. In the frequency domain, the imaging process becomes

$$\mathrm{FT}_{\mathrm{img}}(f_x, f_y) = H_{\mathrm{inc}}(f_x, f_y)\mathrm{FT}_{\mathrm{geom}}(f_x, f_y) \tag{5.17}$$

where $H_{\mathrm{inc}}(f_x, f_y)$ is the normalized transfer function of the optical system.

$$H_{\mathrm{inc}}(f_x, f_y) = \frac{\iint_{-\infty}^{\infty} |h(u, v)|^2 \exp[-\mathrm{j}2\pi(f_x u + f_y v)]\mathrm{d}u\,\mathrm{d}v}{\iint_{-\infty}^{\infty} |h(u, v)|^2\,\mathrm{d}u\,\mathrm{d}v} \tag{5.18}$$

$H_{\mathrm{inc}}(f_x, f_y)$ is known as the optical transfer function (OTF); its modulus $|H_{\mathrm{inc}}(f_x, f_y)|$ is the modulation transfer function (MTF). Because the OTF in (5.17) specifies a weighting factor for the pair of frequencies (f_x, f_y), it transfers the optical information from the geometric object to the image. Similarly, the modulus transfers the modulation (or the contrast) from the geometric object to the image.

The Fourier transform of the impulse response for coherent light, (5.8), is repeated here.

$$H_{\mathrm{coh}}(f_x, f_y) = \iint_{-\infty}^{\infty} h(u, v)\exp\left[-\mathrm{j}2\pi(f_x u + f_y v)\right]\mathrm{d}u\,\mathrm{d}v. \tag{5.19}$$

It can be shown that the incoherent OTF is related to the coherent spectrum of the impulse response by way of an autocorrelation.

$$H_{\mathrm{inc}}(f_x, f_y) = \frac{\iint_{-\infty}^{\infty} H_{\mathrm{coh}}\left(p + \frac{f_x}{2}, q + \frac{f_y}{2}\right)H_{\mathrm{coh}}^{*}\left(p - \frac{f_x}{2}, q - \frac{f_y}{2}\right)\mathrm{d}p\,\mathrm{d}q}{\iint_{-\infty}^{\infty} \left|H_{\mathrm{coh}}(p, q)\right|^2\,\mathrm{d}p\,\mathrm{d}q}. \tag{5.20}$$

The OTF, $H_{inc}(f_x, f_y)$, is the normalized autocorrelation function of the amplitude transfer function $H_{coh}(f_x, f_y)$ with the properties

$$\begin{aligned} H_{inc}(0, 0) &= 1 \\ H_{inc}(-f_x, -f_y) &= H^*_{inc}(f_x, f_y) \\ \left|H_{inc}(f_x, f_y)\right| &\leqslant H_{inc}(0, 0). \end{aligned} \tag{5.21}$$

What do OTFs look like? We will make use of the autocorrelation function above by beginning with the coherent amplitude transfer function (5.11) that is a scaled pupil function.

$$H_{coh}(f_x, f_y) = \mathrm{P}(\lambda z_{img} f_x, \lambda z_{img} f_y). \tag{5.22}$$

Noting that the pupil function (without aberrations) is either 1 or 0, substituting (5.22) into (5.20), we can derive

$$H_{inc}(f_x, f_y) = \frac{\displaystyle\iint_{-\infty}^{\infty} \mathrm{P}\left(x + \frac{\lambda z_{img} f_x}{2}, y + \frac{\lambda z_{img} f_y}{2}\right) \mathrm{P}\left(x - \frac{\lambda z_{img} f_x}{2}, y - \frac{\lambda z_{img} f_y}{2}\right) \mathrm{d}x\, \mathrm{d}y}{\displaystyle\iint_{-\infty}^{\infty} \mathrm{P}^2(x, y) \mathrm{d}x\, \mathrm{d}y}. \tag{5.23}$$

We can see from (5.23) that we have two versions of the pupil moving in opposite directions across the x–y pupil plane. When they overlap, the value in the integrand in the numerator is equal to the area of overlap. By adding the overlap areas via integration, and normalizing by the total area of the pupil (the denominator), the OTF is found. There are many ways to calculate the OTF. For simple geometries such as rectangles, circles and annuli, they are found in closed form, but for most geometric configurations, a computer can make short work of the autocorrelation integral.

For circular pupils, which occur in many physical systems, the OTF can be found in closed form. For a pupil of radius R, the OTF (with radial spatial frequency ρ) is

$$H_{inc}(\rho) = \begin{cases} \dfrac{2}{\pi}\left[\arccos\left(\dfrac{\rho}{2\rho_0}\right) - \dfrac{\rho}{2\rho_0}\sqrt{1 - \left(\dfrac{\rho}{2\rho_0}\right)^2}\right] & \rho \leqslant 2\rho_0 \\ 0 & \text{otherwise} \end{cases} \tag{5.24}$$

where the cutoff frequency is $2\rho_0 = \frac{2R}{\lambda z_{img}}$, which happens to be twice the cutoff frequency of a coherent imaging system, (5.13). This does not imply that an incoherent system has twice the resolving power of a coherent system. As explained very well by Gaskill (1978), the quality of the image is closely related to the spatial frequency content of the object as much as the imaging system, so that direct comparison of coherent or incoherent illumination is not wise.

5.3 Aberrations and their effects

Aberrations can be represented as a deviation from the spherical wavefronts that we have assumed so far. We write this deviation as a phase term applied to the pupil function $\mathrm{P}(x, y)$.

$$\mathrm{P}_{\mathrm{gen}}(x, y) = \mathrm{P}(x, y)\exp\left[\mathrm{j}\frac{2\pi}{\lambda}w(x, y)\right]. \tag{5.25}$$

$\mathrm{P}_{\mathrm{gen}}(x, y)$ is often called the generalized pupil function.

For a coherent imaging system, the generalized pupil function is used in determining the amplitude transfer function. From (5.11),

$$\begin{aligned} H_{\mathrm{coh}}(f_x, f_y) &= \mathrm{P}_{\mathrm{gen}}(\lambda z_{\mathrm{img}} f_x, \lambda z_{\mathrm{img}} f_y) \\ &= \mathrm{P}(\lambda z_{\mathrm{img}} f_x, \lambda z_{\mathrm{img}} f_y)\exp\left[\mathrm{j}\frac{2\pi}{\lambda}w(\lambda z_{\mathrm{img}} f_x, \lambda z_{\mathrm{img}} f_y)\right]. \end{aligned} \tag{5.26}$$

For an incoherent imaging system, we can use the notation of Goodman where the integrals include only the area of overlap of the pupil functions (Goodman 2005). $A(f_x, f_y)$ is defined as the limits where the pupils are separated by the spatial frequencies (f_x, f_y). $A(0, 0)$ represents complete overlap.

$$H_{\mathrm{inc}}(f_x, f_y) = \frac{\iint_{A(f_x, f_y)} \exp\left\{\mathrm{j}\frac{2\pi}{\lambda}\left[w\left(x + \frac{\lambda z_{\mathrm{img}} f_x}{2}, y + \frac{\lambda z_{\mathrm{img}} f_y}{2}\right) - w\left(x - \frac{\lambda z_{\mathrm{img}} f_x}{2}, y - \frac{\lambda z_{\mathrm{img}} f_y}{2}\right)\right]\right\}\mathrm{d}x\,\mathrm{d}y}{\iint_{A(0,0)} \mathrm{d}x\,\mathrm{d}y} \tag{5.27}$$

Because of the complex nature of the generalized pupil function, the OTF can take on negative values and even complex values. With a negative OTF, the image will appear to have a contrast reversal. That is, for those spatial frequencies, bright regions are dark and dark regions appear bright. It can also be shown that the MTF (the squared magnitude of the OTF) with aberrations at (f_x, f_y) is always less than or equal to the unaberrated MTF.

$$\left|H_{\mathrm{inc}}(f_x, f_y)^2\right|_{\text{with aberrations}} \leqslant \left|H_{\mathrm{inc}}(f_x, f_y)^2\right|_{\text{without aberrations}} \tag{5.28}$$

Closed-form solutions for an arbitrary aberration function are usually very difficult to obtain. Even a simple focusing or spherical aberration error is messy. However, computer generated simulations are very common and appear throughout the literature (Mahajan 1983, 1994).

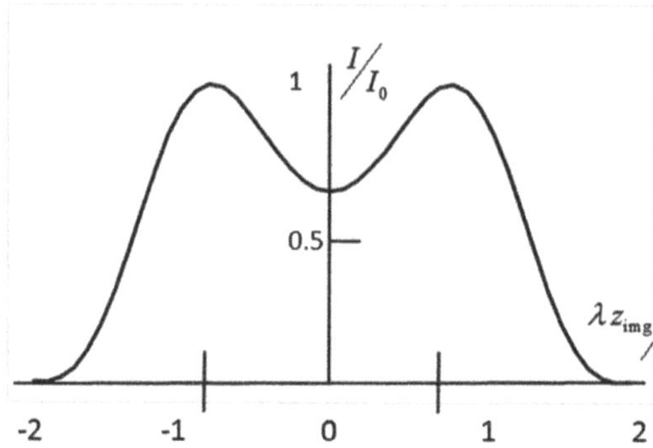

Figure 5.1. Image intensity for two equally bright incoherent point sources separated by the Rayleigh resolution distance.

5.4 Resolution

There are various ways to express the resolution of an imaging system. When resolution is defined as deciding how we see the image of two point sources, we can use the classical Rayleigh criterion. Two incoherent point sources are said to be resolved by a circular aperture when they are separated by the angle (or distance) that has the peak of the Airy disk of one image falling on the first zero of the Airy pattern of the second. The minimum resolvable separation is

$$\widehat{d}_{\text{resolved}} = 0.61\frac{\lambda z_{\text{img}}}{R} \tag{5.29}$$

Figure 5.1 illustrates this. The dip in the intensity at the center is roughly 75% intensity, which is discernible by the eye.

In the case of coherent imaging of two point sources with a circular aperture of radius R, a cross-section of the intensity distribution in the u–v image plane is

$$I\left(\frac{uR}{\lambda z_{\text{img}}}\right) = \left| \frac{2\,J_1\left[\pi\left(\frac{uR}{\lambda z_{\text{img}}} - 0.61\right)\right]}{\pi\left(\frac{uR}{\lambda z_{\text{img}}} - 0.61\right)} + \mathrm{e}^{\mathrm{j}\phi}\,\frac{2\,J_1\left[\pi\left(\frac{uR}{\lambda z_{\text{img}}} + 0.61\right)\right]}{\pi\left(\frac{uR}{\lambda z_{\text{img}}} + 0.61\right)} \right|^2. \tag{5.30}$$

One can see that when the phase difference $\phi = 0$, the intensity distribution is the same as two incoherent point sources. With similarity to Young's double slit experiment, when the phase difference is in opposition, $\phi = \pi$, the intensity reduces to zero, which is plainly visible to the eye.

Many other definitions of resolution are found in the literature. One may argue that the Rayleigh single value definition is too simplistic, and in many cases it is. For example, it has been proposed that the two-dimensional integral of the MTF be used as a figure-of-merit for a measure of resolution.

$$\mathrm{FOM}_{\mathrm{resolution}} = \int_{-\infty}^{\infty} \int \left| H_{\mathrm{inc}}(f_x, f_y)^2 \right| \mathrm{d}f_x \, \mathrm{d}f_y \tag{5.31}$$

This formulation takes into account that a high MTF at all spatial frequencies will mean good resolution and a region where the MTF is low (meaning low contrast in the image) will detract from the measure.

References

Gaskill J D 1978 *Linear Systems, Fourier Transforms, and Optics* (New York: Wiley) p 509

Goodman J W 2005 *Introduction to Fourier Optics* 3rd edn (Englewood, CO: Roberts) p 148

Mahajan V N 1983 Aberrated point-spread functions for rotationally symmetric aberrations *Appl. Opt.* **22** 3035–41

Mahajan V N 1994 Symmetry properties of aberrated point spread functions *J. Opt. Soc. Am.* A **1** 1993–2003

Chapter 6

Enabling applications

It is amazing that a Fourier transform, whose mathematics is well-understood but mechanics can be extremely tedious, is performed at the speed of light by light itself. With most of the fundamentals of Fourier optics understood, over the past half century many applications of the processes have been investigated, developed, and in many cases turned into household products. This chapter will present some of the more established techniques and operations that enable optical system development. The following chapter will show state-of-the-art uses of the Fourier theorem on a beam of light.

6.1 Manipulating light

To take advantage of the seemingly magical ability of light itself to take complex Fourier transforms instantaneously, we need to have the ability to first put the light in a form that needs to be transformed and then capture the light to observe the process. We have to be able to manufacture a beam of light with the proper amplitude and phase to begin the process and capture the amplitude and phase information to see the result.

6.1.1 Refractive and reflective devices

For example, we may want to convert some mechanical or electrical information onto a beam of light. The simplest concept is just blocking out portions of the beam and allowing other specific portions of the beam to propagate.

We can make use of devices such as *liquid crystals* whose optical characteristics can be commanded by electrical signals. Our day-to-day use of liquid crystals in wristwatches, televisions and computer screens proceeds without worrying too much about the underlying physics. However, the rapid developments in miniaturization and cost reduction make it possible to apply them to advanced applications.

doi:10.1088/978-0-750-31056-7ch6

Those everyday devices make use of the process whereby the absorption or phase shift of light within the liquid crystal can be controlled by electric or magnetic fields or by external optical fields, which of course are electromagnetic fields. Details can be found in textbooks and many articles within the literature (Saleh and Teich 1991, Kaneko 1987). The process is generally attributed to the external fields aligning and changing the alignment of the electric dipoles within the molecules of the crystal structure. By varying the orientation of the polarization of incoming light, the liquid crystal can be made (i) to change the state of polarization, (ii) to prohibit passage of the light, (iii) to reflect the light or (iv) to change the local phase of the light. All these forms of manipulation can be called *modulation*, so the term spatial light modulator (SLM) is often used to describe liquid crystals. In some cases, an SLM can be used to convert an incoherent image to a coherent one for later processing.

In addition to liquid crystals, we can modulate light mechanically by reflecting it from a mirror and then modulating the mirror. A *deformable mirror* (DM) can be constructed whereby the surface height and its shape can be electrically or mechanically modified. Many such devices exist on the macroscopic scale (a 10 m DM which is the primary mirror of the Keck telescope (van Dam *et al* 2006)) to miniaturized devices (a 1 cm micromachined DM with thousands of individually controlled regions (Norton *et al* 2009)).

6.1.2 Diffractive devices

The refractive and reflective devices described above can be used in the dynamic sense with some devices possible to modulate light near GHz frequencies. On the other hand, some devices make use of the diffraction of light to modulate the beam. Combining electromagnetic and mechanical drivers, there is a device which makes use of the propagation of sound waves in a medium to modulate light. *Acousto-optic modulators* have a piezoelectric transducer that generates a high frequency sound wave that propagates perpendicular to a beam of light. The variation in the density in the medium results in small variations in the index of refraction of the medium which is viewed as a thin sinusoidal phase grating across the beam. The beam is then sent into various diffractive orders and can be controlled by managing the incident beam angle on the grating inside the acousto-optic modulator.

The miniature deformable mirrors mentioned above use standard photolithography and micromachining techniques but stretch the state-of-the-art by constructing mechanical machines within the structure to bend and move the mirrors. Another method of modulating light is the building of thin complex diffraction gratings on the surface of either reflective or transparent materials. These structures are often constructed for a specific purpose, such as converting a Gaussian beam to a complicated phase and amplitude pattern. Some fabrication processes, such as those used for integrated circuits, build up the diffraction pattern with a series of finite steps of exposing, etching and micromachining. The result of the process is often called *binary optics*. While the initial concept may lend itself to static applications, the combination of two or more diffractive elements, with adjustable positioning between them, can lead to complete dynamic optical systems with structures on the order of a few wavelengths of light (O'Shea *et al* 2003).

6.2 Incoherent optical processing

Coherent imaging systems have one fundamental advantage; there exists a pure frequency plane where the object or image spectrum appeared. For incoherent imaging, the intensity of light is a non-negative real quantity everywhere and there is no way to subtract (or multiply by –1) an intensity pattern without introducing some electronic or mechanical complication.

For example, if we have two transparencies $t_1(\xi, \eta)$ and $t_2(\xi, \eta)$ and simply overlay them, the result that we see is the *product* of the two $t_1(\xi, \eta)t_2(\xi, \eta)$. The trivial case is if $t_1 = 1$ (the lack of a transparency at all) we see only t_2.

Now on the other hand, if we project the two transparencies onto a screen where they overlap, we will see the addition of the two, $t_1(\xi, \eta) + t_2(\xi, \eta)$. The trivial case is where $t_1 = 0$ (the light coming from the first transparency is turned off), we see only t_2. If we now create an optical system where subsequent imaging of the two transparencies is done, the detected light at coordinates (x_d, y_d) appears to be a convolution of the two transparencies. See figure 6.1.

An incoherent source is placed in the front focal plane of lens L_1, which has focal length f. One particular ray emanates from $(-x_s, -y_s)$ on the source. Behind the lens is a transparency with an inverted pattern of the first transparency, i.e. $t_1(-x, -y)$. A second lens L_2, also with focal length f, is placed a distance d behind the first lens and is immediately preceded by the transparency $t_2(x, y)$. The ray from the source and t_1 illuminates t_2 with an intensity proportional to $t_1[-x + (\frac{d}{f})x_s, -y + (\frac{d}{f})y_s]$. The rays are focused onto a detector at coordinates (x_d, y_d). The intensity distribution on the back focal plane of L_2 takes the form of a convolution

$$I(x_d, y_d) = \iint_{-\infty}^{\infty} t_1(x, y)t_2(\xi - x, \eta - y)\mathrm{d}x\,\mathrm{d}y, \tag{6.1}$$

and more specifically for this case,

$$\begin{aligned} I(x_d = x_s, y_d = y_s) &= \iint_{-\infty}^{\infty} t_1\left(\frac{\mathrm{d}x_s}{f} - x, \frac{\mathrm{d}y_s}{f} - y\right)t_2(x, y)\,\mathrm{d}x\,\mathrm{d}y \\ &= t_1(x, y) \otimes t_2(x, y). \end{aligned} \tag{6.2}$$

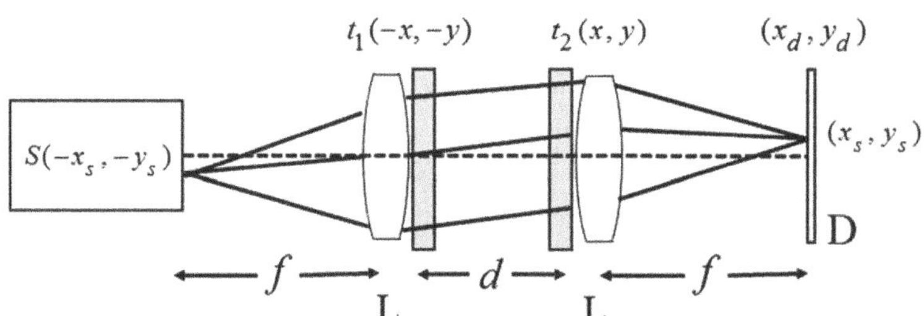

Figure 6.1. System for performing a convolution.

We have now been able to show how multiplication, addition and convolution of two two-dimensional functions can be performed optically.

6.2.1 Spectral imaging

One consequence of imaging and processing with incoherent light is the ability to use the multiple wavelengths in the source to derive spectral information about the object as well as to provide high angular resolution spatial information about the object from various forms of interferometry.

Fourier optics is the common thread between interferometry, high resolution imaging and spectral imaging. From Michelson to Labeyrie (Michelson and Pease 1921, Labeyrie 1975), techniques have evolved with large astronomical installations, such as the Keck interferometer, the Center for High Angular Resolution Astronomy (CHARA), and the Very Large Telescope Interferometer (VLTI), for producing tremendous new data on solar and stellar physics. Similarly, multi-aperture observatories use infrared nulling interferometers (Bracewell and MacPhie 1979, Angel 1990, Angel *et al* 1997) and visible hypertelescopes (Labeyrie 1996) are used in the search for extrasolar planets. Hénault's recent paper joins these various techniques under a single formalism (Hénault 2010).

Beginning with the notation that the on-sky object is represented by angular coordinates $\boldsymbol{u}$, $\boldsymbol{v}$, the spatial coordinates in the entrance pupil plane are $\boldsymbol{x_p}$, $\boldsymbol{y_p}$; the exit pupil plane spatial coordinates are $\boldsymbol{x'_p}$, $\boldsymbol{y'_p}$ and the image detection plane spatial coordinates are $\boldsymbol{x}$, $\boldsymbol{y}$. For multiple apertures, i.e. pupils, the origin of the various coordinate systems is designated by $\boldsymbol{P_n}$, $\boldsymbol{P'_n}$ in the pupils and $\boldsymbol{M'}$ in the image plane. The concept of a multiple aperture imaging system and a multi-arm interferometer can be condensed. The image distribution can be calculated from the integral

$$I(M') = \int_{M'_O}\int O(M'_O)\,|\hat{B}_{D'}(M' - M'_O)|^2 A(M', M'_O)\,\mathrm{d}M'_O. \tag{6.3}$$

The object distribution scaled onto the image plane is $O(M'_O)$. $|\hat{B}_{D'}(M' - M'_O)|^2$ is the diffraction effect from one of the presumably identical subapertures, essentially the PSF. The multiple aperture geometry is combined into

$$A(M', M'_O) = \sum_{n=1}^{N} a_n \exp\left[\mathrm{i}\left(\varphi_n + \frac{2\pi\varsigma_n}{\lambda}\right)\right] \tag{6.4}$$

where a_n and φ_n are the amplitude transmittance and phase difference of the nth subaperture and beam path (arm of the interferometer) and ς_n is the cumulative path length along the nth arm.

Hénault proceeds to analyze a number of configurations under this formalism. In the case of Fizeau-like interferometers, it is shown that the intensity distribution is a convolution product between the scaled object distribution and the system PSF of the multi-aperture optical system. The system PSF here is the PSF of an individual subaperture multiplied by the far-field fringe function (defined below) generated by the geometrical arrangement of subapertures.

For the case of hypertelescopes used for imaging, the convolution relationship no longer holds and the PSF and OTF have a different meaning from the classical single aperture telescope. This is because the hypertelescope has additional optics for combining the beams from the multiple subapertures and rearranging the output pupils to be side-by-side, rather than separated as in a generic multi-aperture telescope.

Another configuration considered by Hénault is a superresolving telescope (a monolithic single pupil optical system with a surface of alternating concentric rings of variable phase differences) which may attain subdiffraction resolution. The image distribution is

$$I(M') = \int_{M_O'} \int O(M_O')\,|\hat{B}_{D'}(M' - M_O')|^2 F\,(M')\,\mathrm{d}M_O'. \tag{6.5}$$

The far-field fringe function is

$$F(M') = \left| \sum_{n=1}^{N} a_n \exp\,[\mathrm{i}\varphi_n]\,\exp\left[\frac{2\pi \mathrm{i}}{\lambda}\frac{\boldsymbol{M'P_n'}}{F'}\right]\right|^2 \tag{6.6}$$

where F' is the distance from the exit pupil to the focal plane. This reduces to the expression

$$I(M') = F(M')\,[O(M_O')*|\hat{B}_{D'}(M')|^2] \tag{6.7}$$

which shows that the image distribution of the superresolving telescope is the constructive and destructive interference of the geometrical subaperture distribution (the fringe function) multiplied by the convolution of the object and the PSF.

6.2.2 Nulling interferometry

An interesting technique, originally proposed by Bracewell, uses multiple apertures (a minimum of two but more if they are used in pairs) to create images where one of the objects in the field-of-view needs to be removed or cut out (Bracewell 1978, Bracewell and MacPhie 1979); see figure 6.2. This requirement occurs in one of the 21st century's most important astronomy problems—imaging extrasolar planets (exoplanets). Because the light from a nearby intragalactic star is much brighter than the reflected light from one of its planets, the starlight will simply mask the planet. The starlight can be up to six orders of magnitude brighter than the planet. Bracewell's technique is now being employed to image exoplanets (Kaltenegger and Fridlund 2005, Lawson *et al* 2007).

Nulling interferometry with two telescopes is easy to explain. The light from the parent star enters the two telescopes on-axis. A phase delay of π radians is placed in one of the beam trains. Thus, because of destructive interference, the on-axis light is nulled. Much dimmer light from the nearby planet enters at a slight angle where the different path lengths through the telescopes added to the constant phase delay results in something other than total destructive interference and the planet's light

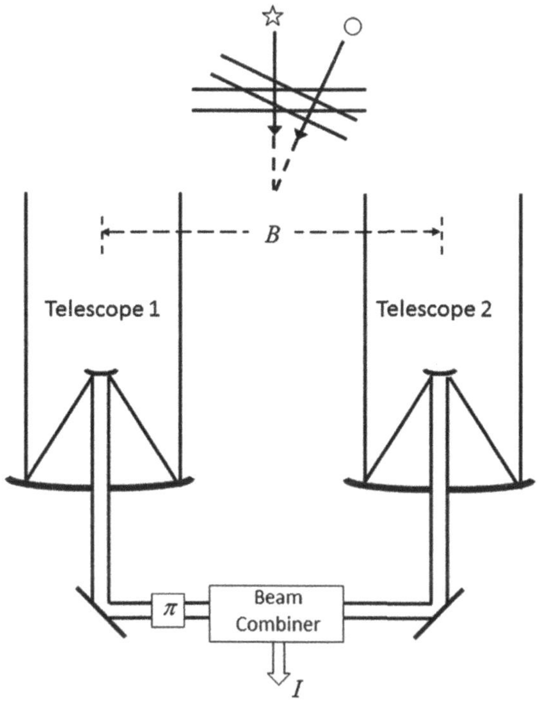

Figure 6.2. A stellar nulling interferometer.

reaches the detector. Laurent *et al* (2006) describe many of the engineering details for implementation of multi-aperture nulling interferometers.

6.2.3 Snapshot spectral imaging

Spectral imaging can take many forms and have various names. Imaging spectrometry, imaging spectroscopy, hyperspectral imaging and multispectral imaging all provide a means of taking a two-dimensional image with each image region, a pixel at (x, y), stretched in a third dimension into its spectrum. The resulting (x, y, λ) data are typically called a data cube. Scanning spectral imaging is time-sequential two-dimensional slices of the cube. Snapshot spectral imaging is a simultaneous measurement of all two-dimensional spatial elements that are recombined into a cube in post-processing.

With the development of one-dimensional linear sensor arrays in the 1970s, some early researchers developed the point scanning spectrometer, which used the array to capture the spectrum of a single spatial point and then make a two-dimensional spatial scan with one or more galvanometer (galvo) mirrors (Potter 1972). With two-dimensional detector arrays, the field evolved rapidly in the 1980s. In the pushbroom spectrometer, the two-dimensional array scans one line across the spatial dimension and fills the other dimension with spectral information (Gehm *et al* 2008). The other spatial dimension is achieved via the moving platform or the scene being scanned with a galvo mirror.

Spectral information is captured via a filter wheel or an electrically tunable filter such as a Fabry–Perot etalon (Atherton *et al* 1981, Antila *et al* 2012). An imaging Fourier transform spectrometer scans one mirror of a Michelson interferometer to obtain multiple OPD values which is the Fourier equivalent of a tunable filter camera (Potter 1972, Descour 1996). Alternatively, one can use the birefringent Fourier-transform imaging spectrometer developed by Harvey and Fletcher-Holmes (2004).

Ferrec *et al* (2011b) describe the development of a high étendue static Fourier transform spectral imager that consists of a spectral filter in the image plane. Rather than narrowband filters with a variable wavenumber (Lucey *et al* 2008, Ma *et al* 2007), Ferrec's design has filters that are sinusoidal in the spectral domain with a variable period. Two reflecting dihedrons in the legs of the Michelson interferometer move orthogonal to one another. One moves along the beam path of one leg, changing the OPD, while the other moves laterally, thereby creating a lateral shearing interferometer (Ferrec *et al* 2011a). This fixed combination of an imaging system and an interferometer allows for a large throughput (étendue). In recent years many variations of these techniques have been developed that make use of new electro-optic technologies and improvements (Eismann 2012, Prieto-Blanco *et al* 2008, Fernandez *et al* 2007, Gehm *et al* 2008, Chakrabarti *et al* 2012).

Versions of snapshot imaging spectrometers have been explored for the better part of a century (Bowen 1938). Many of the early systems suffered from manufacturing difficulties and alignment constraints. Over the years, many tradeoffs were made in terms of spatial versus spectral resolution and manufacturing difficulty and cost. The concept of microslicing allowed the measurement of many spatial elements in the image with a reduction in spectral sampling (Content *et al* 2003). The use of coherent fiber bundles and lenslet arrays for feeding the spectrometers have seen development from the 1980s to the present day (Bland-Hawthorn *et al* 2011, Lee *et al* 2001, Allington-Smith and Content 1998, Leon-Saval *et al* 2010). In astronomy and Earth-sensing applications, there has been a constant advance of technology and invention in the past decade or so. The tunable echelle imager (Baldry and Bland-Hawthorn 2000) and the use of spectrally resolving detector arrays (Bilbro 2001) continue to evolve (Buchsbaum and Morris 2003, Shrestha *et al* 2011, Gupta *et al* 2011, Parrein *et al* 2009).

The image-replicating imaging spectrometer is based on Lyot's invention of the tunable filter to perform wavelength separation. This approach makes use of the high efficiency of modern detectors. It has been shown to operate with up to eight spectral bands (Gorman *et al* 2010) and is probably physically limited to 16 spectral bands (Wong *et al* 2011).

A few other snapshot spectral imagers are described briefly in Hagen and Kudenov (2013). These include the coded aperture snapshot spectral imager (CASSI; Gehm *et al* 2007), the image mapping spectrometer (IMS; Gao *et al* 2009), the snapshot hyperspectral imaging Fourier transform spectrometer (SHIFT; Kudenov and Dereniak 2010), and the multispectral Sagnac interferometer (MSI; Kudenov *et al* 2010). Implementation of any of these methods includes judicious selection of the detector arrays and efficiency of the optics in dividing the light into elements of the data cube.

6.3 Optical coherence tomography

In a typical interferometer, interference fringes only appear if the OPD between the two legs is less than the coherence distance of the source. In conventional interferometry with long coherence lengths (laser interferometry), interference of light occurs over a distance of meters. Optical coherence tomography (OCT) is a method that employs partially coherent (short coherence length) light with optical detection and processing to produce three-dimensional images of transparent optically scattering media such as biological tissue. OCT is based on low coherence interferometry. (Riederer 2000, Born and Wolf 2000, Fercher *et al* 1988.)

In OCT, the interference is shortened to a distance of micrometers with the use of broadband light sources. White light is also a broadband source with lower power (Huang *et al* 1991).

OCT with micrometer resolution and cross-sectional imaging capabilities has become a prominent biomedical tissue-imaging technique. It is particularly suited to ophthalmic applications and other tissue imaging requiring micrometer resolution and millimeter penetration depth. Starting from white-light interferometry for *in vivo* ocular eye measurements, imaging of biological tissue, especially of the human eye, was investigated by multiple groups worldwide (Fercher and Roth 1986, Huang *et al* 1991, Fercher *et al* 1993, Ortiz *et al* 2013).

6.3.1 Time-domain OCT

Light in an OCT system is broken into two arms—a sample arm (containing the item of interest) and a reference arm (usually a mirror). The combination of reflected light from the sample arm and reference light from the reference arm gives rise to an interference pattern, but only if light from both arms has traveled an optical distance less than a coherence length. By scanning the mirror in the reference arm, a reflectivity profile of the sample can be obtained which results in time-domain OCT (TD-OCT). Areas of the sample that reflect back a lot of light will create greater interference than areas that do not. Any light that is outside the short coherence length will not interfere. This reflectivity profile, called an A-scan, contains information about the spatial dimensions and location of structures within the item of interest. A cross-sectional tomograph (B-scan) may be achieved by laterally combining a series of these axial depth scans (A-scans).

The principal OCT is white light or low coherence interferometry. The optical set-up typically consists of a Michelson type interferometer (figure 6.3) with a low coherence, broad bandwidth light source. Light is split into and recombined from the reference and sample arms, respectively.

This interference is called auto-correlation in a symmetric interferometer (both arms have the same reflectivity), or cross-correlation in the common case. The envelope of this modulation changes as path length difference is varied, where the peak of the envelope corresponds to path length matching. The interference of two partially coherent light beams, I, can be expressed in terms of the source intensity I_s

$$I = k_1 I_s + k_2 I_s + 2\sqrt{k_1 I_s k_2 I_s}\,\mathrm{Re}[\gamma(\tau)] \tag{6.8}$$

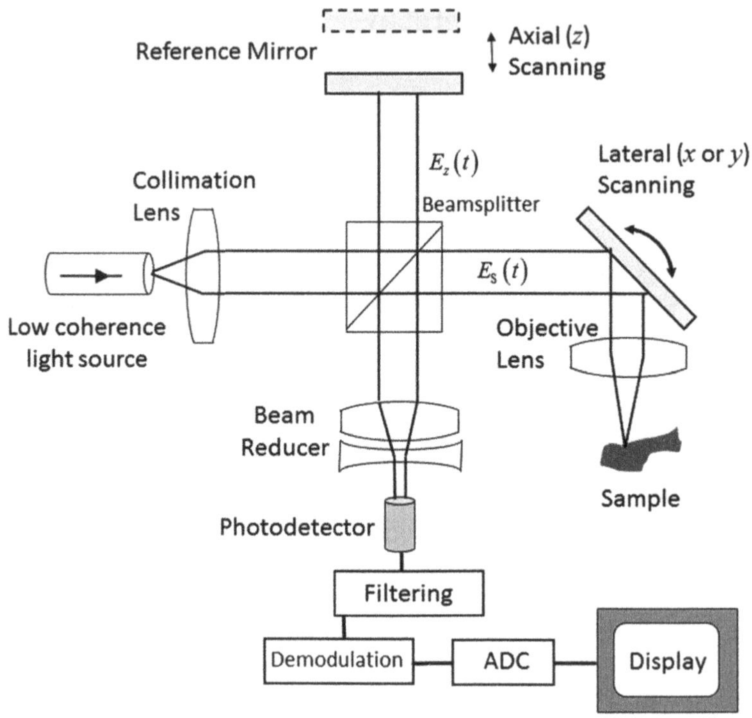

Figure 6.3. Typical optical set-up of single point OCT. Scanning the light beam on a sample enables non-invasive cross-sectional imaging up to 3 mm in depth with micrometer resolution.

where $k_1 + k_2 < 1$ is the interferometer beamsplitting ratio and $\gamma(\tau)$ is the complex degree of coherence. For reference arm scan time τ, center frequency ν_0 and bandwidth $\Delta\nu$ of the source, the complex degree of coherence can be represented by

$$\gamma(\tau) = \exp\left[-\left(\frac{\pi\Delta\nu\tau}{2\sqrt{\ln 2}}\right)^2\right]\exp(-j2\pi\nu_0\tau). \tag{6.9}$$

The Gaussian envelope is amplitude modulated by an optical carrier. The peak of this envelope represents the location of the test sample microstructure, with an amplitude dependent on the reflectivity of the surface. The optical carrier is due to the Doppler effect, resulting from scanning one arm of the interferometer, and the frequency of this modulation is controlled by the speed of scanning. Therefore translating one arm of the interferometer has two functions; depth scanning and a Doppler-shifted optical carrier are accomplished by path length variation. In OCT, the Doppler-shifted optical carrier has a frequency expressed as

$$f_{\text{Doppler}} = \frac{2\nu_0\,\text{v}_s}{c} \tag{6.10}$$

where v_s is the scanning velocity of the path length variation and c is the speed of light.

The axial and lateral resolutions of OCT are different. Axial resolution r_{Axial} is the same as the coherence length of the source.

$$r_{\text{Axial}} = \frac{2\ln 2}{\pi} \cdot \frac{\lambda_0^2}{\Delta\lambda} \approx 0.44\frac{\lambda_0^2}{\Delta\lambda} \tag{6.11}$$

Lateral resolution is defined classically by the Rayleigh criterion.

6.3.2 Frequency domain OCT

In frequency domain OCT (FD-OCT) the broadband interference is acquired with spectrally separated detectors, either by encoding the optical frequency in time with a spectrally scanning source or with a dispersive detector, such as a grating and a linear detector array (Schmitt 1999). The depth scan can be immediately calculated by a Fourier transform from the acquired spectra, without movement of the reference arm (Zhang *et al* 2013b).

There are hybrid techniques used by researchers in the field. Spatially-encoded frequency domain OCT (SEFD-OCT) extracts spectral information by distributing different optical frequencies onto a detector stripe (line-array CCD or CMOS) via a dispersive element. Time-encoded frequency domain OCT (TEFD-OCT) encodes the spectral components, not by spatial separation but by time separation. A comparison of TD-OCT and FD-OCT has been performed (Leitgeb 2003).

6.3.3 Industrial and medical applications

For decades, OCT has been used for industrial quality control and maintenance (Takada *et al* 1987) and high resolution metrology (Dufour *et al* 2005, Tomczewski *et al* 2013). Advances in the technology are rapidly being adopted by the medical community. Where one can show that OCT is superior, either faster, less costly or less invasive, with adequate training, the clinic adopts it (Marshall 2014). The biggest growth areas are ophthalmology (Keane *et al* 2012, Swanson *et al* 1993), dermatology, dentistry and cardiovascular endoscopy (Bezerra *et al* 2009).

6.4 Optical testing

6.4.1 Collimation testing

One of the simplest optical tests of a laser, other than simply observing that it is on, is the determination that the beam is collimated, neither converging nor diverging. Rather than measuring beam diameters at various planes along the beam, a shear plate can be inserted into the beam at one plane. Using the principle of the shearing interferometer (Mallick and Malacara 2007), the transparent plate with very flat surfaces and a slight wedge is inserted at an angle of 45° to the beam direction, see figure 6.4. When a plane wave is incident on the plate, two reflected wavefronts result. These are separated laterally (sheared) because of the plate thickness and angularity due to the wedge. With plane wavefronts incident, the area of overlap between the two reflected beams will show fringes when projected on a screen.

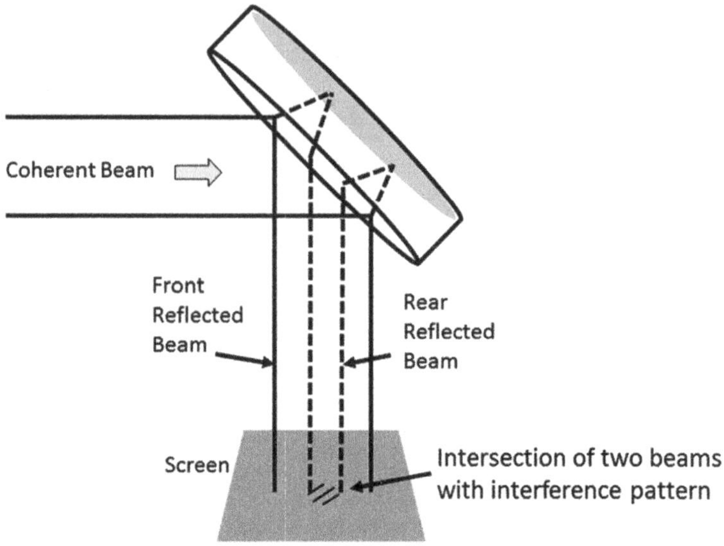

Figure 6.4. Use of a shear plate to test collimation.

These fringes will be perpendicular to the wedge orientation. If they are straight and oriented in the direction of propagation, the beam is collimated. If the straight fringes are rotated, or their slope changes, the beam shows some vergence. If the fringes are not straight, some aberrations other than vergence are present.

Another method for determining collimation is based on the Talbot effect (Silva 1971). Described in Dhanotia and Prakash (2011), the automated collimation testing is built on the effect that a grating of pitch p is placed in a plane-parallel wavefront, a self-image of the grating is observed at a distance

$$d = \frac{2p^2 m}{\lambda} \tag{6.12}$$

where m is an integer. If a second grating of the same pitch is placed at the first Talbot image, a moiré pattern is formed. By filtering the Fourier transform of the moiré fringes, the wavefront slope information can be recovered.

6.4.2 Alignment

An extension of the moiré fringe analysis for precise optical alignment for nano-lithography has been proposed (Mühlberger *et al* 2007). Noting that specific moiré patterns can appear only when the gratings are properly aligned, the patterns can be used to determine alignment of the gratings used in lithography (King and Berry 1972). A thorough theoretical analysis can be found in the literature (Zhou *et al* 2008).

6.4.3 Wavefront testing

Beyond simply determining collimation, lateral shearing interferometers have been in use for wavefront measurement for decades (Wyant 1973, Joenathan *et al* 1984).

When the beam under test is not collimated, holographic elements can be added to collect the data (Malacara and Mallick 1976). Typically, the analysis is performed by time stepping, but recently a combination of multiplexing holographic lenses and calculating a spatial Fourier transform has been shown to lead to the two-dimensional wavefront analysis from a single frame (Joenathan *et al* 2012, 2013). In this system, a holographic lens (fabricated from the interference between a wave from a point source and one from a collimated laser beam) is made with three exposures. One exposure has no shear, one has x-shear, and the third has y-shear. When the test beam with wavefront $w(x, y)$ is transmitted through the holographic lens, the fringe pattern appears as

$$f(x, y) = I(x, y)\left\{1 + M\exp\left[\mathrm{j}\left(\frac{2\pi}{\lambda}\frac{\partial w}{\partial x}\Delta x + 2\pi\frac{\alpha}{\lambda}x\right)n\right]\right\}. \tag{6.13}$$

The beam intensity is $I(x, y)$, the fringe modulation is M, Δx is the shear, α is the angle between the two beams that produce the shear and n is the diffraction order ($n = 1$ for first-order diffraction). The wavefront slope in the x-direction can be recovered by Fourier transforming the fringe pattern, filtering out the zero-order spectrum, centering one of the first-order spectra and inverse transforming. The shear fringes from the phase portion of the modified fringe pattern,

$$f'(x, y) \propto I(x, y)M\,\exp\left[\mathrm{j}\left(\frac{2\pi}{\lambda}\frac{\partial w}{\partial x}\Delta x\right)n\right] \tag{6.14}$$

can be used to uncover the wavefront slope $\frac{\partial w}{\partial x}$ and the second pattern can be used to recover $\frac{\partial w}{\partial y}$.

In addition to interferometric methods, one can use Shack–Hartmann wavefront sensors based on point sources (Cao and Yu 1994, Platt and Shack 2001) or based on resolved scenes (Poyneer 2003).

6.4.4 Non-destructive materials testing

Using the principles of Fourier optics, two-dimensional imaging can be used to study various parameters of the object of interest. For example, flatness, edge parallelism and surface quality can be analyzed to nanometer precision (Dhanasekar and Ramamoorthy 2008). Fourier optics far-field optical analyzers and interferometry have been shown to be useful in characterizing diffractive optical elements (Huang *et al* 2006), analyzing laser-damaged surfaces (Theocaris 1997) and investigating highly asymmetrical objects (Tamulevicius *et al* 2006). Recently, there has been work focused on investigating all of these parameters using one robust edge-quality analyzer with a transmissive optics design. See figure 6.5.

As a demonstration of analyzing a submillimeter-thick silicon wafer bar, the reflected light from the edge sample is Fourier transformed by a lens. While a perfect

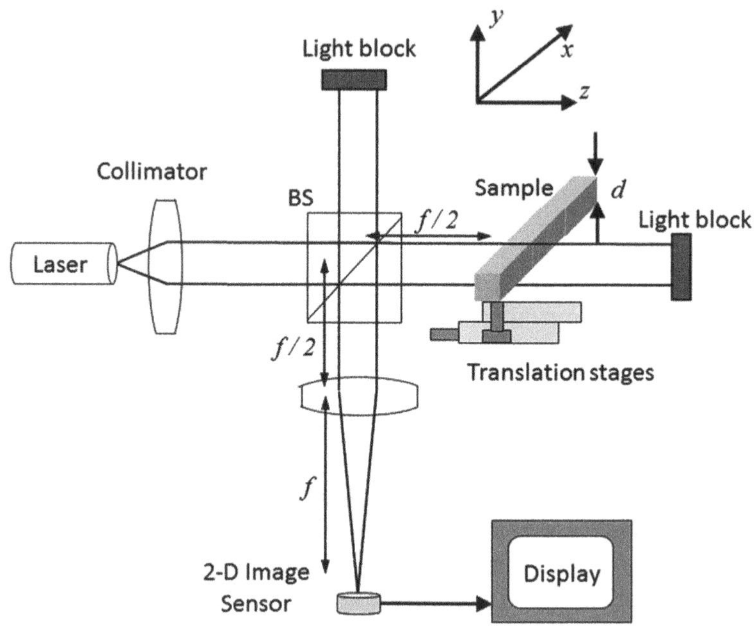

Figure 6.5. An edge quality analyzer with transmissive optics.

rectangular edge $U(y) = \frac{1}{d}\mathrm{rect}(\frac{y}{d})$ would create a sinc function in the Fourier plane $U(\frac{y}{\lambda f}) = \mathrm{sinc}(\frac{yd}{\lambda f})$, the intensity in the two-dimensional image would be

$$I \propto \left|\mathrm{sinc}\left(\frac{yd}{\lambda f}\right)\right|^2. \tag{6.15}$$

When the edge has defects, the side lobes in the Fourier plane can be depressed and shifted. By analyzing the precise position and amplitude of the side lobes in the Fourier plane, the edge of the wafer bar can be characterized quantitatively (Sumriddetchkajorn and Chaitavon 2011).

6.5 Optical information processing

Optical information processing, that is, using photons for computation, has existed since the 1870s (Abbe 1873). The most common implementations are the Fourier domain processor where the filter is in the Fourier domain and the joint transform processor where the filter is in the spatial domain. Although these optical processors are easily (or sometimes with great frustration) designed for a specific purpose, it is difficult to make one programmable or able to make decisions like their electrical equivalent. Advances in hybrid optical–electronic computing have been used for scene and pattern recognition, inspection of materials and synthetic aperture radar (SAR) (Yu 2002). Researchers have combined optical processing with electronic optimization algorithms such as simulated annealing (Yin *et al* 1996) and neural networks (Psaltis and Farhat 1985). The combination of parallel (optical) computing and serial (electronic) computing has recently seen rapid advances with some of our most stringent and demanding problems (Dolev and Oltean 2012).

6.5.1 Superresolution

Superresolution (SR) is a class of techniques that enhance the resolution of an imaging system (Simpkins and Stevenson 2012). In some techniques, such as optical SR, the diffraction limit of systems is transcended, while in others, such as geometrical SR, the resolution of digital imaging sensors is enhanced. While the physics of diffraction cannot be neglected, it was shown by Lukosz decades ago (Lukosz 1967, 1997) that some degrees of freedom of an imaging system could be enhanced at the expense of others. For example, the space bandwidth product could be enlarged if the field-of-view were reduced (Cox and Sheppard 1986). An explosion of research in the 1960s created the field which now extends into nanolithography (Maria *et al* 2004), digital holography (Paturzo *et al* 2008) and microscopy (Yamaguchi *et al* 2001). No laws of physics are broken when a spatial frequency band beyond the cut-off spatial frequency is swapped for one inside it. This has long been implemented in dark-field microscopy (Gustaffsson 2000). Using *a priori* information is also applicable to superresolving systems. Resolution involves the distinction of whether a target is single or double, ordinarily requiring separable peaks in the image. When a target is *known to be single*, its location can be determined with higher precision than the image width by finding the centroid (center of gravity) of its image light distribution.

Research that has been pursued in recent years includes the recording of a multiplexed hologram where the individual components each contain information about a different spatial frequency region of the diffracted object spectrum. The individual diffracted orders must not overlap so that the recovered digital Fourier transform can recover the individual passbands. For a practical off-axis implementation of the technique, some information in each passband must be discarded (the overlap) which necessarily restricts the field-of-view. By introducing a diffraction grating in the imaging arm of the interferometer, the previously discarded passbands can be recovered and directed toward the detector for recording and analysis (Granero *et al* 2009). This has the effect of increasing the effective aperture and providing SR.

Another technique to achieve SR is phase-shifting lithography, which uses a π phase-shifting mask near the object which creates destructive interference near the edges and thus enhances them (Levenson *et al* 1982). Advancement in calculating (Hsu and Su 2007) and producing the phase-shifting patterns (Paturzo *et al* 2008) has been outlined and demonstrated experimentally (Hsu *et al* 2011).

6.6 Spatial filtering and propagation

6.6.1 Fractional Fourier transform

The fractional Fourier transform (FRFT) is a family of linear transformations that generalize the Fourier transform. It is essentially a Fourier transform to the *n*th power where *n* is not necessarily an integer. By its nature it can transform a function to an intermediate domain between time and frequency or between space and spatial frequency. The FRFT can be used to define fractional convolutions and correlations and extend Shannon's sampling theorem in the fractional Fourier domain (Bhandari and Marziliano 2010).

Beginning with the Fourier transform and its inverse

$$\mathrm{FT}[g(t)] = G(\nu) = \int_{-\infty}^{\infty} g(t)\exp[-\mathrm{j}2\pi\nu t]\,\mathrm{d}t$$
$$\mathrm{FT}^{-1}[G(\nu)] = g(t) = \int_{-\infty}^{\infty} G(\nu)\exp[+\mathrm{j}2\pi\nu t]\,\mathrm{d}\nu \tag{6.16}$$

examine FT^n where $\mathrm{FT}^n(g) = \mathrm{FT}[\mathrm{FT}^{n-1}(g)]$ and $\mathrm{FT}^{-n}(g) = [\mathrm{FT}^{-1}(g)]^n$ when n is a non-negative integer. Also $\mathrm{FT}^0(g) = g$. The sequence is finite and FT is a four-period automorph, i.e. $\mathrm{FT}^4(g) = g$. Note that $\mathrm{FT}^2[g(t)] = g(-t)$, reversing time or inverting space. This can be extended to the non-integer n values.

If a regular Fourier transform converts (transforms) a time-domain signal to a frequency domain spectrum, a fractional Fourier transform converts a signal or a spectrum into a domain between time and frequency. The transform, dependent upon the value of n, simply rotates the distribution in the time frequency plane through an angle $\theta = \frac{n\pi}{2}$ (Almeida 1994). The expression for the FRFT becomes

$$\mathrm{FT}^n[g(t)] = \int g(t)K_n(t;\,\nu)\,\mathrm{d}t \tag{6.17}$$

where the transform kernel is

$$K_n(t;\,\nu) = \sqrt{1 - \mathrm{j}\cot\theta}\,\exp\left[\mathrm{j}\pi\left(\frac{\nu^2 + t^2}{\tan\theta} - \frac{2t\nu}{\sin\theta}\right)\right]. \tag{6.18}$$

Applications include temporal and spatial filtering where one can filter both time and frequency information simultaneously with a single transform and inverse (Bhandari and Marziliano 2010). Zhang *et al* (2013a) extend the discrete FRFT methods of Yeh and Pei to derive an algorithm for a parallel computation method (Yeh and Pei 2003). With simultaneous transforms with multiple angles, an efficient large-scale signal processing integration is possible.

Another adaption of the FRFT is the introduction of random kernels in the FRFT. By randomly modulating the phase angle θ, a signal or an image or any high-dimensional object can be encrypted and decrypted using an optical system. These random Fourier transforms can be implemented optically using both gradient index materials (Mendlovic and Ozaktas 1993) and lenses (Lohmann 1993). Liu and Liu (2007) show a system (figure 6.6) with an input and an output in contact with random phase masks $P(x)$ and $P^*(f_x)$ with the relationship between the input function $g_{\mathrm{in}}(x)$ and the output transform $g_{\mathrm{out}}(f_x)$ being

$$g_{\mathrm{out}}(f_x) = P^*(f_x)\mathrm{FT}^n[g_{\mathrm{in}}(x)P(x)]. \tag{6.19}$$

There is a further generalization of the FRFT that has been shown to be a method of propagating a variety of beams (such as Gaussian, cos-Gaussian, flat-top, Hermite

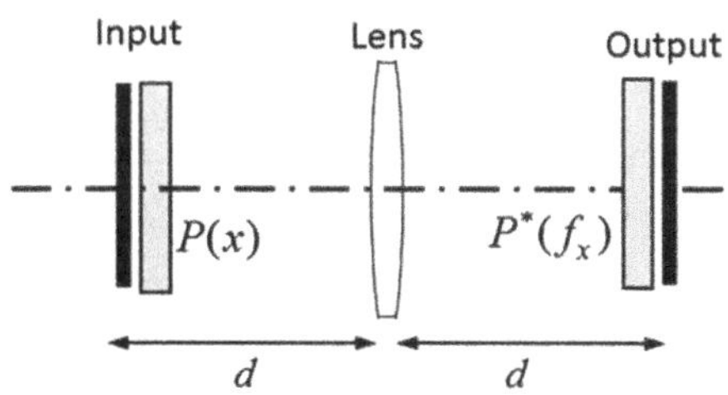

Figure 6.6. Optical implementation of the random fractional Fourier transform.

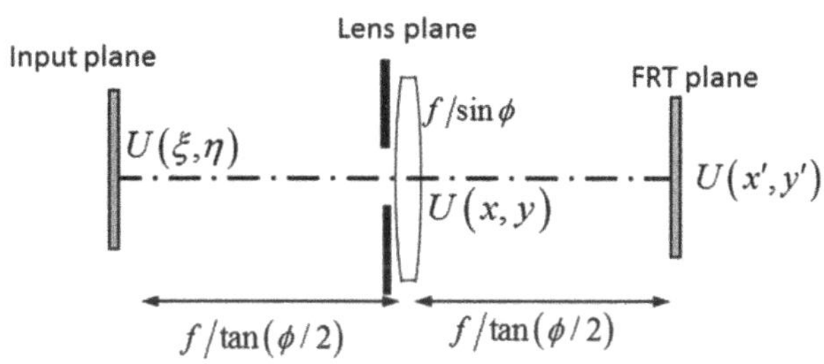

Figure 6.7. System for a fractional Fourier transform propagation and modification of various beams.

Gaussian, higher order annular Gaussian and others). Using a system that looks like figure 6.7, an FRFT can be used to propagate a generalized beam and is a convenient method for beam shaping and other all-optical modifications (Zhao and Cai 2010). Using the diffraction integral written in terms of the ABCD matrix (Collins 1970), the field at the truncated lens plane $U(x, y)$ is related to the input field $U(\xi, \eta)$ by

$$U(x, y) = \frac{\mathrm{j}}{\lambda B_1} \exp\left[-\mathrm{j}kf \tan\left(\frac{\phi}{2}\right)\right] \iint_{-\infty}^{\infty} U(\xi, \eta) \\ \times \exp\left\{-\frac{\mathrm{j}\pi}{\lambda B_1}[A_1(\xi^2 + \eta^2) - 2(x\xi + y\eta) + D_1(x^2 + y^2)]\right\} \mathrm{d}\xi\, \mathrm{d}\eta. \quad (6.20)$$

The angle $\phi = n\frac{\pi}{2}$ and n is the fractional order of the Fourier transform. The matrix elements represent the ABCD elements in propagation from the input to the lens plane.

$$A_1 = 1 \qquad B_1 = f \tan\left(\frac{\phi}{2}\right) \qquad C_1 = 0 \qquad D_1 = 1. \quad (6.21)$$

After further propagation to the FRFT plane, the field is given by

$$U(x', y') = \frac{\mathrm{j}}{\lambda B_2} \exp\left[-\mathrm{j}\frac{2\pi}{\lambda} f \tan\left(\frac{\phi}{2}\right)\right] \iint_{-\infty}^{\infty} H(x, y)\, U(x, y, 0) \\ \times \exp\left\{-\frac{\mathrm{j}\pi}{\lambda B_2}[A_2(x^2 + y^2) - 2(xx' + yy') + D_2(x'^2 + y'^2)]\right\} \mathrm{d}x\, \mathrm{d}y. \quad (6.22)$$

The ABCD elements for the second propagation are

$$A_2 = \cos\phi \qquad B_2 = f\tan\left(\frac{\phi}{2}\right) \qquad C_2 = -\sin\left(\frac{\phi}{f}\right) \qquad D_2 = 1. \quad (6.23)$$

The beauty of the formulation is that the input field can be a general-type beam consisting of Hermite polynomials with any specified order and a set of complex numbers V_{xl} and V_{yl} representing the physical location and phase rotation of the field. We can represent the field by summing N beams in this general notation

$$U(\xi, \eta, 0) = \sum_{l=1}^{N} A_l \exp(-j\phi_l)\, H_{nl}(a_{xl}\xi + b_{xl}\eta)\, H_{ml}(a_{yl}\xi + b_{yl}\eta)$$
$$\times \exp\left[\frac{-1}{\lambda}(a_{xl}\xi^2 + b_{xl}\eta^2)\right] \exp[-j(V_{xl}\xi + V_{xl}\eta)]. \quad (6.24)$$

Setting the various orders of the Hermite polynomials corresponds to possibly an infinite number of possibilities for the researcher to explore. For example, a Gaussian beam is selected by choosing $N = 1$, $l = 1$, $A_l = 0.5$, $nl = ml = V_{xl} = V_{yl} = 0$ (Zhao and Cai 2010).

This method of calculating the output of the FRFT optical system has been extended to elliptical Gaussian beams (Cai and Lin 2003a), Lorentz–Gaussian beams (Zhou 2009a), Ince–Gaussian beams (Zhou 2009b), Bessel–Gaussian beams (Tang *et al* 2012) and partially coherent beams (Cai and Lin 2003b). Furthermore, in general, Fresnel integrals corresponding to different distances can be interpreted as scaled FRFTs observed on spherical surfaces. To accurately calculate the Fresnel integrals numerically, Ozaktas *et al* (2012) show how to interpret the lateral spacing requirements in relation to the longitudinal propagation distance.

6.6.2 Propagation studies

Using variations of the Fresnel and Fraunhofer diffraction integrals to numerically calculate optical wave fields along propagation paths has been around since the 1960s and the advent of high-speed computers. The angular spectrum method calculates the optical field along a grid of points at points along the path. First, the complex input field is produced on a (normally) square grid with each point represented by an amplitude and phase at that point. A two-dimensional Fourier transform, typically using the fast Fourier transform (FFT) algorithm, decomposes the field into a two-dimensional angular spectrum of component plane waves each traveling in a unique direction. Then the spectrum is modified by a propagation term accounting for the phase change along the propagation path. Finally, an inverse two-dimensional Fourier transform results in a representation of the field in the output plane. Volumes have been written about the techniques, modifications and especially the limitations of the method (Delen and Hooker 1998), but new ideas are constantly appearing in the literature (Cincotti 2011).

The principal difficulty in the method occurs where the sampling grid becomes very large and bogs down the computational process. This occurs for many well-known scenarios: (a) very large beams with phase profiles on a small scale (e.g. beams propagating through atmospheric turbulence); (b) wildly non-symmetric beams (e.g. Airy beams); and (c) fairly large beams that are propagated through a tight focus where the Fresnel approximation does not apply.

6.6.2.1 Large beams

This book will briefly address some recent research applicable to these difficulties. For problem (a) above, we examine the space–bandwidth product limitation. From chapter 4 we note that Fourier transforming a function of small spatial extent leads to a spectrum of large frequency extent. Similarly, transforming a large spatial extent leads to a small frequency extent. Thus, if we have a large optical beam with either small amplitude and phase variations or worse yet, a hard aperture with infinite spatial frequency, we have to have a very large number of grid points in the input plane so that there is enough resolution in the frequency plane to detect the higher frequencies in the input. Subsequent propagation exposes the problem further. Some of the solutions to the computation problem are driven by the calculation of computer generated holograms and the like. While band-limiting the angular spectrum using an elliptically shaped filter in the Fourier plane can limit some of the information, it has been shown that it is within the aliasing limits and information transfer requirements for computational or digital holography (Matsushima and Shimobaba 2009). In some cases, large aperture approximations can be applied to smaller optical systems, thereby simplifying the analysis (Porter *et al* 2013).

6.6.2.2 Non-symmetric beams

For problem (b) above, recent research discusses algorithms for calculating the properties of various forms of Airy beams. Airy beams (Berry and Balazs 1979) have the property of intensity invariance and a parabolic propagation path, very useful in optical tweezing, for example (Broky *et al* 2008). When combined with a spiral phase plate, the beam has a singularity in the center and orbital angular momentum but still maintains the same intensity distribution. The Airy beam is defined to be

$$A(x, y) = \exp\left[\frac{a}{w}(x + y)\right]\mathrm{Ai}\left(\frac{x}{w}\right)\mathrm{Ai}\left(\frac{y}{w}\right) \tag{6.25}$$

where the curvature a and the width w are arbitrary transverse scales defining the exponential truncation. The Airy function $\mathrm{Ai}(q)$ is defined by (Berry and Balazs 1979)

$$\mathrm{Ai}(q) = \frac{1}{2\pi}\int_{-\infty}^{\infty}\exp\left[\mathrm{j}\left(\frac{t^3}{3} + qt\right)\right]\mathrm{d}t. \tag{6.26}$$

Sending the beam through the vortex (spiral) phase plate, the vortex Airy beam becomes

$$U(x, y) = \frac{[(x - x_d) + \mathrm{j}(y - y_d)]^m}{[(x - x_d)^2 + (y - y_d)^2]^{\frac{m}{2}}}A(x, y). \tag{6.27}$$

The beam propagates along the z-axis and rotates about the point (x_d, y_d) with an m-order vortex (Chu 2012).

In a mathematical sense, Airy beams have infinite energy (the integrals extend to infinity). That is why they can propagate without exhibiting some of the properties of Fresnel diffraction. In reality, finite energy beams must exist and they are described as incomplete Airy beams because they have a sharp spectral cutoff. The incomplete Airy beam is described by

$$\mathrm{Ai}^{\mathrm{inc}}_{M,W}(q) = \frac{1}{2\pi}\int_{M-W}^{M+W} \exp\left[\mathrm{j}\left(\frac{t^3}{3} + qt\right)\right]\mathrm{d}t \tag{6.28}$$

where M is the center of the integration window and W is the width. The total intensity in the beam is

$$\int_{-\infty}^{\infty} \left|\mathrm{Ai}^{\mathrm{inc}}_{M,W}(q)\right|^2 \,\mathrm{d}q = \frac{W}{\pi}. \tag{6.29}$$

The expectation value of the transverse position of the beam as a function of propagation distance is

$$\langle q \rangle = \frac{-W^2}{3} - M^2 + 2M\,z \tag{6.30}$$

which means the beam center is notably linear in z.

6.6.2.3 Beams near focus

For problem (c) above, we can examine the propagation modeling at the National Ignition Facilty (NIF) in Livermore, CA, USA. NIF is a high energy laser system intended to achieve thermonuclear ignition. It consists of 192 independent beams aimed onto a 2 mm target. The total energy is 1.8 MJ in 20 ns pulses. Each of the 192 beams has hundreds of optical components with an optical path near 4.5 km (Haynam *et al* 2007). When using Fourier-based beam models of large beam trains, the Talanov algorithm is used, to avoid the accumulation of off-axis errors (Talanov 1970). Revisiting the difficulties mentioned above, because the beams experience an intensity singularity near focus, major modifications to the angular spectrum method are necessary to avoid the quadratic phase associated with focus and the transverse grid limitations that result in loss of spatial resolution and aliasing.

The Talanov algorithm separates the quadratic phase term from the field amplitude expression and constantly updates the effective focal distance $f(z) = f_0 - z$ and scales the grid by the factor $f_0(f_0 - z)^{-1}$. This method allows the beam to propagate almost to the focus, or to propagate it through focus to after focus. However, if the propagation step ends inside the Rayleigh range, $z_{\mathrm{Rayleigh}} = \frac{\pi w_0^2}{\lambda}$ (w_0 is the beam waist), the distribution is essentially a point and all information is lost for the next propagation step. The approach chosen to avoid this dilemma is by using the

Talanov algorithm very close to focus and then modifying the parameters near the focal plane. This approach maintains a larger beam size near the focus with a more gradual shrinking of the grid. Only small focal length shifts are required and the quadratic phase is well-resolved (Feigenbaum *et al* 2013).

6.7 Spatial filters

Spatial light modulation is often separated into amplitude or phase modulation with separate processes, such as two spatial light modulators (Neto *et al* 1996), or common path interferometry (Reichelt *et al* 2012). One method demonstrated recently performs simultaneous phase and amplitude modulation using only a single phase-only spatial light modulator by adding Fourier filtering (Sarkadi *et al* 2013). Using a 4-f system such as that shown in figure 6.8, a phase-only modulation is applied to the input beam. The beam propagates to a Fourier-transforming lens. At the Fourier plane, a spatial filter is applied. The beam then is propagated through the second Fourier-transforming lens and a complex modulated wavefront appears at the output plane.

6.8 Optical correlators

An optical correlator is a device used for target tracking and identification using the Fourier transforming property of a lens and signal comparisons. The correlator has an input signal which is multiplied by some filter in the Fourier domain. An example filter is the matched filter which uses the cross correlation of the two signals.

One can create the optical cross correlation $c(x, y)$ of two two-dimensional signals $g(x, y)$ with $h(x, y)$ using a convex lens. The cross correlation, denoted by

$$c(x, y) = g(x, y) \otimes h^*(x, y) \tag{6.31}$$

is represented in the Fourier domain as

$$C(\xi, \eta) = G(\xi, \eta)H^*(\xi, \eta) \tag{6.32}$$

where the capital letters denote the Fourier transform of the lower case letter. The cross-correlation is calculated by inverse Fourier transforming the result.

A convex lens of focal length f produces a Fourier transform of the coherent distribution $g(x, y)$ located a distance f in front of the lens. In a 4-f correlator, the Fourier plane is in between the two lenses. The complex signal $G(\xi, \eta)$ is multiplied

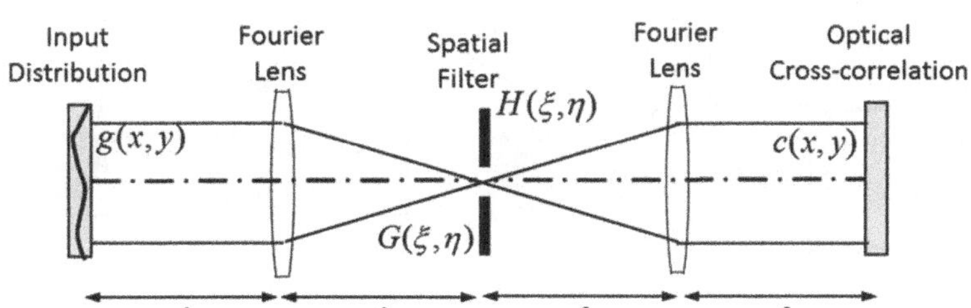

Figure 6.8. 4-f system for optical cross-correlation.

by the complex filter $H(\xi, \eta)$ and then inverse transformed by the second lens into a version of $c(x, y)$. The filter can be as simple as an optical transparency for a static system, or it can be optically modulated by one or more spatial light modulators.

There are two principal types of optical correlators. The two types are the matched filter, based on the VanderLugt filter (VanderLugt 1964) and the joint transform correlator (Weaver and Goodman 1966).

6.8.1 The VanderLugt matched filter

Let us first discuss the matched filter, its implementation and its information processing. The process of generating the matched filter $H(\xi, \eta)$ begins with reference signal $h(x, y)$. The filter can be a complete complex filter

$$H(\xi, \eta) = \mathrm{FT}[h(x, y)] \tag{6.33}$$

or a phase-only filter (Horner and Gianino 1984)

$$H(\xi, \eta) = \frac{\mathrm{FT}[h(x, y)]}{|\mathrm{FT}[h(x, y)]|}. \tag{6.34}$$

The process for the matched filter involves taking the individual Fourier transforms of the signal $g(x, y)$ and reference $h(x, y)$ images, taking their product and then Fourier transforming the result. See figure 6.9. Since 1964, many improvements have been made to implement the filter using spatial light modulators, but the process of optically Fourier transforming the signals remains the basis for the extremely fast technique. See figure 6.10.

The complex filter can be constructed in a number of ways (Goodman 2005, section 8.4.1). However, it should always take the form of a transmissive scaled multiple of $H(\xi, \eta)$, the Fourier transform of the impulse response $h(x, y)$. The final lens of a 4-f system or alternative configuration will result in a field at the output plane (coordinates x, y) which is a sum of four terms.

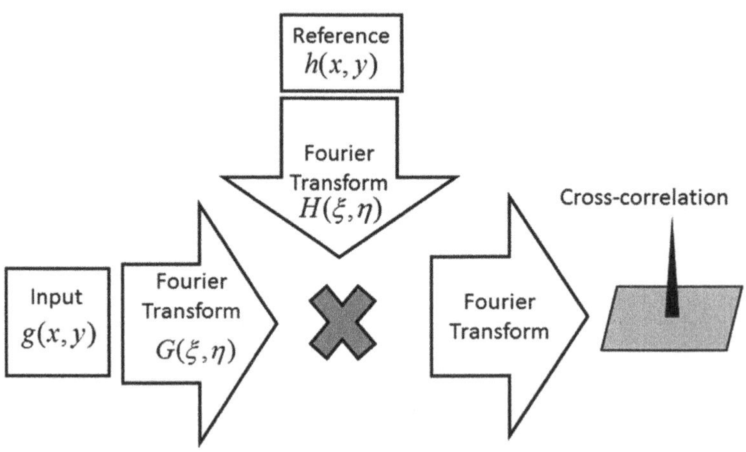

Figure 6.9. Flow chart for the system to perform an optical cross-correlation.

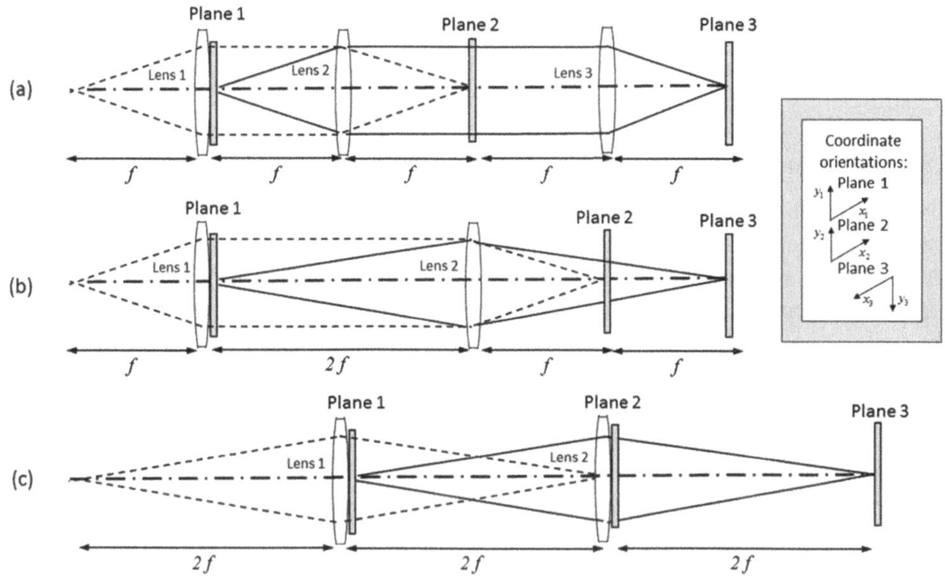

Figure 6.10. Architectures for coherent optical information processing.

$$U(x, y) \propto r_0 g(x, y) + \frac{1}{\lambda^2 f^2}[h(x, y) \otimes h^*(-x, -y) \otimes g(x, y)]$$
$$+ \frac{r_0}{\lambda f}[h(x, y) \otimes g(x, y) \otimes \delta(x, y + \alpha\lambda f)]$$
$$+ \frac{r_0}{\lambda f}[h^*(-x, -y) \otimes g(x, y) \otimes \delta(x, y - \alpha\lambda f)]. \tag{6.35}$$

What appears are four unique distributions in the x–y output plane. r_0 is the amplitude of the reference wave that formed the filter and α is the spatial frequency associated with the relative tilt between the two beams $h(x, y)$ and $g(x, y)$; the first two terms are centered at the origin of the x–y output plane. The third term is centered at $(0, -\alpha\lambda f)$ in the output plane and is proportional to the convolution of h and g.

$$h(x, y) \otimes g(x, y) \otimes \delta(x, y + \alpha\lambda f) = \iint_{-\infty}^{\infty} h(x - \xi, y + \alpha\lambda f - \eta) g(\xi, \eta) \mathrm{d}\xi\, \mathrm{d}\eta \tag{6.36}$$

The fourth term is centered at $(0, \alpha\lambda f)$, symmetrically opposed to the position of the third term, and takes the form of the cross-correlation of g and h.

$$h^*(-x, -y) \otimes g(x, y) \otimes \delta(x, y - \alpha\lambda f)$$
$$= \iint_{-\infty}^{\infty} g(\xi, \eta) h^*(\xi - x, \eta - y + \alpha\lambda f) \mathrm{d}\xi\, \mathrm{d}\eta \tag{6.37}$$

If $\alpha > \frac{1}{\lambda f}(\frac{3W_h}{2} + W_g)$, where W_h and W_g are the widths of the individual distributions h and g, respectively, the convolution and cross-correlation distributions are sufficiently separated to avoid aliasing or crosstalk.

6.8.2 Joint transform correlator

The joint transform correlator differs significantly from the matched filter process, since no reference filter is required. Instead, the fields h and g are aligned and displayed alongside each other, separated laterally by s, on the input plane, typically a spatial light modulator. The two images are then Fourier transformed together producing the joint power spectrum (JPS). The JPS is then Fourier transformed itself, producing an output of symmetrically identical pairs of correlation peaks. See figure 6.11.

The field transmitted through the front focal plane is

$$U_1(x_1, y_1) = h\left(x_1, y_1 - \frac{s}{2}\right) + g\left(x_1, y_1 + \frac{s}{2}\right) \tag{6.38}$$

and the Fourier transform of this field is found at the back focal plane of the lens.

$$U_2(x_2, y_2) = \frac{1}{\lambda f}H\left(\frac{x_2}{\lambda f}, \frac{y_2}{\lambda f}\right)e^{-j\pi y_2 s/\lambda f} + \frac{1}{\lambda f}G\left(\frac{x_2}{\lambda f}, \frac{y_2}{\lambda f}\right)e^{-j\pi y_2 s/\lambda f}. \tag{6.39}$$

After adjusting for nonlinearities in the recording and playback (detector and SLM), and then retransmitting through the lens, the output field in spatial coordinates (x, y) is

$$\begin{aligned} U(x, y) = \; & \frac{1}{\lambda f}[h(x, y) \otimes h^*(-x, -y) + g(x, y) \otimes g^*(-x, -y)] \\ & + h(x, y) \otimes g^*(-x, -y) \otimes \delta(x, y - s) \\ & + h^*(-x, -y) \otimes g(x, y) \otimes \delta(x, y + s). \end{aligned} \tag{6.40}$$

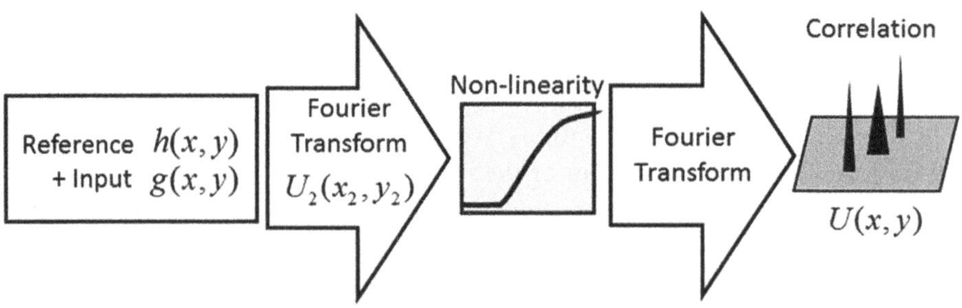

Figure 6.11. The joint transform correlator.

The third and fourth terms are both crosscorrelations of h and g.

$$h(x, y) \otimes g^*(-x, -y) \otimes \delta(x, y - s) = \iint_{-\infty}^{\infty} h(\xi, \eta) g^*(\xi - x, \eta - y + s) \mathrm{d}\xi \, \mathrm{d}\eta \tag{6.41}$$

$$h^*(-x, -y) \otimes g(x, y) \otimes \delta(x, y + s) = \iint_{-\infty}^{\infty} g(\xi, \eta) h^*(\xi - x, \eta - y - s) \mathrm{d}\xi \, \mathrm{d}\eta. \tag{6.42}$$

To get a convolution, it is necessary to start with a mirror reflection of either h or g, but not both. The result will be two distributions that represent the convolution of the functions. If the functions h and g are not real, a complex conjugate, in addition to the reflection, must be used.

References

Abbe E 1873 Contributions to the theory of the microscope and microscopic perception *Archiv. Mikrosk. Anatomie* **9** 413–63

Allington-Smith J and Content R 1998 Sampling and background subtraction in fiber-lenslet integral field spectrographs *Publ. Astron. Soc. Pac.* **110** 1216–34

Almeida L B 1994 The fractional Fourier transform and time frequency representations *IEEE Trans. Signal Process.* **42** 3084–91

Angel J R 1990 Use of a 16 m telescope to detect earthlike planets *Proc. Workshop on the Next Generation Space Telescope* ed P Bely and C J Burrows (Baltimore, MD: Space Telescope Science Institute) pp 81–94

Angel J R, Burge J H and Woolf N J 1997 Detection and spectroscopy of exo-planets like Earth *Proc. SPIE* **2871** 516–9

Antila J, Mannila R, Kantojärvi U, Holmlund C, Rissanen A, Näkki I, Ollila J and Saari H 2012 Spectral imaging device based on tuneable MEMS Fabry–Perot interferometer *Proc. SPIE* **8374** 83740F

Atherton P D, Reay N K, Ring J and Hicks T R 1981 Tunable Fabry–Perot filters *Opt. Eng.* **20** 806–14

Baldry I K and Bland-Hawthorn J 2000 A tunable echelle imager *Publ. Astron. Soc. Pac.* **112** 1112–20

Bezerra H G 2009 Intracoronary optical coherence tomography: a comprehensive review *JACC: Cardiovascular Interventions* **2** 1035–46

Berry M V and Balazs N L 1979 Non-spreading wave packets *Am. J. Phys.* **47** 264

Bhandari A and Marziliano P 2010 Sampling and reconstruction of sparse signals in fractional Fourier domain *IEEE Signal Process. Lett.* **17** 221–4

Bilbro G L 2001 Technology options for multi-spectral infrared camera *North Carolina State University Technical Report* AD-A396058

Bland-Hawthorn J *et al* 2011 Hexabundles: imaging fiber arrays for low-light astronomical applications *Opt. Express* **19** 2649–61

Born M and Wolf E 2000 *Principles of Optics: Electromagnetic Theory of Propagation, Interference and Diffraction of Light* (Cambridge: Cambridge University Press)

Bowen I S 1938 The image slicer, a device for reducing loss of light at slit of stellar spectrograph *Astrophys. J.* **88** 113–24

Bracewell R N 1978 Detecting nonsolar planets by spinning infrared interferometer *Nature* **274** 780–1

Bracewell R N and MacPhie R H 1979 Searching for non-solar planets *Icarus* **38** 136–47

Broky J, Siviloglou G A, Dogariu A and Christodoulides D N 2008 Self-healing properties of optical Airy beams *Opt. Express* **16** 12880–91

Buchsbaum P E and Morris M J 2003 Method for making monolithic patterned dichroic filter detector arrays for spectroscopic imaging *US Patent* No. 6638668 B2

Cai Y and Lin Q 2003a Fractional Fourier transform for elliptical Gaussian beam *Opt. Commun.* **217** 7–13

Cai Y and Lin Q 2003b Transformation and spectrum properties of partially coherent beams in the fractional Fourier transform plane *J. Opt. Soc. Am.* **20** 1528–36

Cao G and Yu X 1994 Accuracy analysis of a Hartmann–Shack wavefront sensor operated with a faint object *Opt. Eng.* **33** 2331–5

Chakrabarti S, Jokiaho O-P, Baumgardner J, Cook T, Martel J and Galand M 2012 High-throughput and multislit imaging spectrograph for extended sources *Opt. Eng.* **51** 013003

Chu X 2012 Propagation of an Airy beam with a spiral phase *Opt. Lett.* **37** 5202–4

Cincotti G 2011 Generalized fiber Fourier optics *Opt. Lett.* **36** 2321–3

Collins S A 1970 Lens-system diffraction integral written in terms of matrix optics *J. Opt. Soc. Am.* **60** 1168–77

Content R, Morris S and Dubbeldam M 2003 Microslices and low cost spectrographs for million element integral field spectroscopy *Proc. SPIE* **4842** 174–82

Cox I J and Sheppard C J R 1986 Information capacity and resolution in an optical system *J. Opt. Soc. Am.* A **3** 1152–8

Delen N and Hooker B 1998 Free-space beam propagation between arbitrarily oriented planes based on full diffraction theory: a fast Fourier transform approach *J. Opt. Soc. Am.* A **15** 857–67

Descour M R 1996 The throughput advantage in imaging Fourier transform spectrometers *Proc. SPIE* **2819** 285–90

Dhanasekar B and Ramamoorthy B 2008 Digital speckle interferometry for assessment of surface roughness *Opt. Lasers Eng.* **46** 272–80

Dhanotia J and Prakash S 2011 Automated collimation testing by incorporating the Fourier transform method in Talbot interferometry *Appl. Opt.* **50** 1446–52

Dolev S and Oltean M (ed) 2012 *Optical Supercomputing, 4th Int. OSC Workshop (Bertinoro, Italy, July 2012)* (Berlin: Springer)

Dufour M L, Lamouche G, Detalle V, Gauthier B and Sammut P 2005 Low-coherence interferometry, an advanced technique for optical metrology in industry *Insight* **47** 216–9

Eismann M T 2012 *Hyperspectral Remote Sensing* (Bellingham, WA: SPIE)

Feigenbaum E, Sacks R A, McCandless K P and MacGowan B J 2013 Algorithm for Fourier propagation through the near-focal region *Appl. Opt.* **52** 5030–5

Fercher A F, Hitzenberger C K, Drexler W, Kamp G and Sattmann H 1993 *In vivo* optical coherence tomography *Am. J. Ophthalmol.* **116** 113–4

Fercher A F, Mengedoht K and Werner W 1988 Eye-length measurement by interferometry with partially coherent light *Opt. Lett.* **13** 186–8

Fercher A F and Roth E 1986 Ophthalmic laser interferometry *Proc. SPIE* **658** 48–51

Fernandez C, Guenther B D, Gehm M E, Brady D J and Sullivan M E 2007 Longwave infrared (LWIR) coded aperture dispersive spectrometer *Opt. Express* **15** 5742–53

Ferrec Y, Taboury J, Sauer H and Chavel P 2011a Compactness of lateral shearing interferometers *Appl. Opt.* **50** 4656–63

Ferrec Y, Taboury J, Sauer H, Chavel P, Fournet P, Coudrain C, Deschamps J and Primot J 2011b Experimental results from an airborne static Fourier transform imaging spectrometer *Appl. Opt.* **50** 5894–904

Gao L, Kester R T and Tkaczyk T S 2009 Compact image slicing spectrometer (ISS) for hyperspectral fluorescence microscopy *Opt. Express* **17** 12293–308

Gehm M E, Kim M S, Fernandez C and Brady D J 2008 High-throughput multiplexed push-broom hyperspectral microscopy *Opt. Express* **16** 11032–43

Gehm M E, John R, Brady D J, Willett R M and Schulz T J 2007 Single-shot compressive spectral imaging with a dual-disperser architecture *Opt. Express* **15** 14013–27

Goodman J W 2005 *Introduction to Fourier Optics* 3rd edn (Greenwood Village, CO: Roberts)

Gorman A, Fletcher-Holmes D W and Harvey A R 2010 Generalization of the Lyot filter and its application to snapshot spectral imaging *Opt. Express* **18** 5602–9

Granero L, Micó V, Zalevsky Z and García J 2009 Superresolution imaging method using phase-shifting digital lensless Fourier holography *Opt. Express* **17** 15008–20

Gupta N, Ashe P R and Tan A 2011 Miniature snapshot multispectral imager *Opt. Eng.* **50** 033203

Gustaffsson M 2000 Surpassing the lateral resolution limit by a factor of two using structured illumination microscopy *J. Microsc.* **198** 82–87

Hagen N and Kudenov M W 2013 Review of snapshot spectral imaging technologies *Opt. Eng.* **52** 090901

Harvey A R and Fletcher-Holmes D W 2004 Birefringent Fourier-transform imaging spectrometer *Opt. Express* **12** 5368–74

Haynam C A *et al* 2007 National ignition facility laser performance status *Appl. Opt.* **46** 3276–303

Hénault F 2010 Simple Fourier optics formalism for high-angular-resolution systems and nulling interferometry *J. Opt. Soc. Am.* A **27** 435–49

Horner J L and Gianino P D 1984 Phase-only matched filtering *Appl. Opt.* **23** 812–6

Hsu W-F, Chen Y and Su Y 2011 Implementation of phase-shift patterns using a holographic projection system with phase-only diffractive optical elements *Appl. Opt.* **50** 3646–52

Hsu W-F and Su Y-H 2007 Implementation of far-field phase shift lithography using diffractive optical elements *Proc. SPIE* **6832** 68320J

Huang J G, Christian J M and McDonald G S 2006 Fresnel diffraction and fractal patterns from polygonal apertures *J. Opt. Soc. Am.* A **23** 2768–74

Huang D *et al* 1991 Optical coherence tomography *Science* **254** 1178–81

Joenathan C, Bernal A and Sirohi R S 2013 Spatially multiplexed *x–y* lateral shear interferometer with varying shears using holographic lens and spatial Fourier transform *Appl. Opt.* **52** 5570–6

Joenathan C, Mohanty R K and Sirohi R S 1984 Lateral shear interferometry with holo shear lens *Opt. Commun.* **52** 153–6

Joenathan C, Pedrini G, Alekseenko I and Osten W 2012 Novel and simple lateral shear interferometer with holographic lens and spatial Fourier transform *Opt. Eng.* **51** 075601

Kaltenegger L and Fridlund M 2005 The Darwin mission: search for extra-solar planets *Adv. Space Res.* **36** 1114–22

Kaneko E 1987 *Liquid Crystal TV Displays: Principles and Applications of Liquid Crystal Displays* (Tokyo: KTK Scientific)

Keane P A, Patel P J, Liakopoulos S, Heussen F M, Sadda S R and Tufail A 2012 Evaluation of age-related macular degeneration with optical coherence tomography *Surveys Opthalmol.* **57** 389–414

King M C and Berry D H 1972 Photolithographic mask alignmernt using moiré techniques *J. Vac. Sci. Technol.* B **11** 2455–8

Kudenov M W and Dereniak E L 2010 Compact snapshot birefringent imaging Fourier transform spectrometer *Proc. SPIE* **7812** 781206

Kudenov M W, Jungwirth M E L, Dereniak E L and Gerhart G R 2010 White-light Sagnac interferometer for snapshot multispectral imaging *Appl. Opt.* **49** 4067–75

Labeyrie A 1975 Interference fringes obtained on Vega with two optical telescopes *Astrophys. J.* **196** L71–5

Labeyrie A 1996 Resolved imaging of extra-solar planets with future 10–100 km optical interferometric arrays *Astron. Astrophys. Suppl. Ser.* **118** 517–24

Laurent F, Hénault F, Ferruit P, Prieto E, Robert D, Renault E, Dubois J-P and Bacon R 2006 CRAL activities on advanced image slicers: optical design, manufacturing, assembly, integration and testing *New Astron. Rev.* **50** 346–50

Lawson P R, Lay O P, Johnston K J and Beichman C A (ed) 2007 *Jet Propulsion Laboratory, California Institute of Technology, TPF-I Science Working Group Report* JPL Pub. 07–1

Lee D, Haynes R, Ren D and Allington-Smith J 2001 Characterization of lenslet arrays for astronomical spectroscopy *Publ. Astron. Soc. Pac.* **113** 1406–19

Leitgeb R, Hitzenberger C K and Fercher A F 2003 Performance of Fourier domain vs time domain optical coherence tomography *Opt. Express* **11** 889–94

Leon-Saval S G, Argyros A and Bland-Hawthorn J 2010 Photonic lanterns: a study of light propagation in multimode to single-mode converters *Opt. Express* **18** 8430–9

Levenson M D, Viswanathan N S and Simpson R A 1982 Improving resolution in photolithography with a phase-shifting mask *IEEE Trans. Electron Dev.* **29** 1828–36

Liu Z and Liu S 2007 Random fractional Fourier transform *Opt. Lett.* **32** 2088–90

Lohmann A W 1993 Image rotation, Wigner rotation, and the fractional Fourier transform *J. Opt. Soc. Am.* A **10** 2181–6

Lucey P G, Horton K A and Williams T 2008 Performance of a long-wave infrared hyperspectral imager using a Sagnac interferometer and an uncooled microbolometer array *Appl. Opt.* **47** F107–13

Lukosz W 1966 Optical systems with resolving powers exceeding the classical limit *J. Opt. Soc. Am.* **56** 1463–72

Lukosz W 1967 Optical systems with resolving powers exceeding the classical limit: II *J. Opt. Soc. Am.* **57** 932–41

Ma X, Yang J, Qiao W and Xue B 2007 An improved Fourier-based sub-pixel image registration algorithm for raw image sequence of LASIS *Proc. SPIE* **6623** 66230A

Malacara D and Mallick S 1976 Holographic lateral shear interferometer *Appl. Opt.* **15** 2695–7

Mallick S and Malacara D 2007 Common-path interferometers *Optical Shop Testing* ed D Malacara (New York: Wiley) pp 97–120

Maria J, Jeon S and Rogers J A 2004 Nanopatterning with conformable phase masks *J. Photochem. Photobiol.* **A166** 149–54

Marshall L S 2014 OCT Technologies, applications showing great promise *Photonics Showcase (March 2014)* (Pittsfield, MA: Laurin)

Matsushima K and Shimobaba T 2009 Band-limited angular spectrum method for numerical simulation of free-space propagation in far and near fields *Opt. Express* **17** 19662–73

Mendlovic D and Ozaktas H M 1993 Fractional Fourier transforms and their optical implementation: I *J. Opt. Soc. Am.* A **10** 1875–81

Michelson A A and Pease F G 1921 Measurement of the diameter of α Orionis with the interferometer *Astrophys. J.* **53** 249–59

Mühlberger M *et al* 2007 A moiré method for high accuracy alignment in nanoimprint lithography *Microelectron. Eng.* **84** 925–7
Neto L G, Roberge D and Sheng Y 1992 Full-range continuous, complex modulation by the use of two coupled-mode liquid-crystal televisions *Appl. Opt.* **31** 163–5
Norton A, Evans J W, Gavel D, Dillon D, Palmer D, Macintosh B, Morzinski K and Cornelissen S 2009 Preliminary characterization of Boston Micromachines' 4096-actuator deformable mirror *Proc. SPIE* **7209** 72090I
Ortiz S, Pérez-Merino P and Marcos S 2013 Eye biometry using quantitative 3-D OCT *Biomed. Opt. Express* **4** 387–9
O'Shea D C, Suleski T J, Kathman A D and Prather D W 2003 *Diffractive Optics: Design, Fabrication and Test* (*SPIE Tutorial Texts in Optical Engineering TT62*) (Bellingham, WA: SPIE)
Ozaktas H M, Arık S Ö and Coşkun T 2012 Fundamental structure of Fresnel diffraction: Longitudinal uniformity with respect to fractional Fourier order *Opt. Lett.* **37** 103–5
Parrein P, Moussy N, Poupinet L and Gidon P 2009 Multilayer structure for a spectral imaging sensor *Appl. Opt.* **48** 653–7
Paturzo M 2008 Flexible coherent diffraction lithography by tunable phase arrays in lithium niobate crystals *Opt. Commun.* **281** 1950–3
Paturzo M, Merola F, Grilli S, De Nicola S, Finizio A and Ferraro P 2008 Super-resolution in digital holography by a two-dimensional dynamic phase grating *Opt. Express* **16** 17107–18
Platt B and Shack S 2001 History and principles of Shack–Hartmann wavefront sensing *J. Refractive Surg.* **17** S573–7
Porter C, Gordeyev S and Jumper E 2013 Large-aperture approximation for not-so-large apertures *Opt. Eng.* **52** 071417
Potter A E 1972 Multispectral imaging system *US Patent* No. 3702735
Poyneer L A 2003 Scene-based Shack–Hartmann wave-front sensing: analysis and simulation *Appl. Opt.* **42** 5807–15
Prieto-Blanco X, Montero-Orille C, Couce B and de la Fuente R 2008 Optical configurations for imaging spectrometers *Comput. Intell. Remote Sensing* **133** 1–25
Psaltis D and Farhat N 1985 Optical information processing based on an associative-memory model of neural nets with thresholding and feedback *Opt. Lett.* **10** 98–100
Reichelt S, Häussler R, Fütterer G, Leister N, Kato H, Usukura N and Kanbayashi Y 2012 Full-range, complex spatial light modulator for real-time holography *Opt. Lett.* **37** 1955–7
Riederer S J 2000 Current technical development of magnetic resonance imaging *IEEE Eng. Med. Biol.* **19** 34–41
Saleh B E A and Teich M C 1991 *Fundamentals of Photonics* (New York: Wiley)
Sarkadi T, Kettinger Á and Koppa P 2013 Spatial filters for complex wavefront modulation *Appl. Opt.* **52** 5449–54
Schmitt J M 1999 Optical coherence tomography (OCT): a review *IEEE J. Quantum Electron.* **5** 1205
Shrestha R, Hardeberg J Y and Khan R 2011 Spatial arrangement of color filter array for multispectral image acquisition *Proc. SPIE* **7875** 787503
Silva D E 1971 A simple interferometric technique for collimation testing *Appl. Opt.* **10** 1980–2
Simpkins J and Stevenson R L 2012 An introduction to super-resolution imaging *Mathematical Optics: Classical, Quantum and Computational Methods* ed V Lakshminarayanan, M Calvo and T Alieva (Boca Raton, FL: CRC Press) pp 539–64
Sumriddetchkajorn S and Chaitavon K 2008 A Fourier-optics-based non-invasive and vibration-insensitive micron-size object analyzer for quality control assessment *Proc. SPIE* **6995** 699511

Sumriddetchkajorn S and Chaitavon K 2011 Simultaneous analysis of edge quality parameters for submillimeter-thick silicon wafer bar with Fourier optics *Opt. Eng.* **50** 023605-1-7

Swanson E A, Izatt J A, Hee M R, Huang D, Lin C P, Schuman J S, Puliafito C A and Fujimoto J G 1993 *In vivo* retinal imaging by optical coherence tomography *Opt. Lett.* **18** 1864–6

Takada K, Yokohama I, Chida K and Noda J 1987 New measurement system for fault location in optical waveguide devices based on an interferometric technique *Appl. Opt.* **26** 1603–06

Talanov V I 1970 Focusing of light in cubic media *JETP Lett.* **11** 199–201

Tamulevicius S, Janusas G, Guobiene A, Palevicius A, Ostasevicius V and Andrulevicius M 2006 Optical characterization of diffractive optical elements replicated by polymers *J. Microlithogr. Microfabr. Microsys.* **5** 013004

Tang B, Jiang C and Zhu H 2012 Propagation of Bessel–Gaussian beams through a double-apertured fractional Fourier transform optical system *J. Opt. Soc. Am.* A **29** 1728–33

Theocaris P S 1997 Light scattering from laser-damaged speckled surfaces *Appl. Opt.* **36** 8755–84

Tomczewski S, Pakula A, Van Erps J, Thienpont H and Salbut L 2013 Low-coherence interferometry with polynomial interpolation on Compute Unified Device Architecture-enabled graphics processing units *Opt. Eng.* **52** 094105

van Dam M A *et al* 2006 The W M Keck Observatory laser guide star adaptive optics system: performance characterization *Pub. Astro. Soc. Pac.* **118** 840

VanderLugt A 1964 Signal detection by complex spatial filtering *IEEE Trans. Inform. Theory* **10** 139–45

Weaver C J and Goodman J W 1966 A technique for optically convolving two functions *Appl. Opt.* **5** 1248–9

Wong G, Pilkington R and Harvey A R 2011 Achromatization of Wollaston polarizing beam splitters *Opt. Lett.* **36** 1332–4

Wyant J C 1973 Double frequency grating lateral shear interferometer *Appl. Opt.* **12** 2057–60

Yamaguchi I, Kato J-I, Ohta S and Mizuno J 2001 Image formation in phase-shifting digital holography and applications to microscopy *Appl. Opt.* **40** 6177–85

Yeh M H and Pei S C 2003 A method for the discrete fractional Fourier transform computation *IEEE Trans. Signal Process.* **51** 889–91

Yin S, Hudson T D, McMillen D K, Lu M, Chen C and Yu F T S 1996 Design of a bipolar composite filter using simulated annealing algorithm *Opt. Lett.* **20** 1409–11

Yu F T S 2002 Legacy of optical information processing *Opt. Eng.* **41** 139–44

Zhang F, Tao R and Wang Y 2013a Discrete fractional Fourier transform computation by adaptive method *Opt. Eng.* **52** 068202

Zhang X, Gao W, Bian H, Chen C and Liao J 2013b Self-spectral calibration for spectral domain optical coherence tomography *Opt. Eng.* **52** 063603

Zhao C and Cai Y 2010 Propagation of a general-type beam through a truncated fractional Fourier transform optical system *J. Opt. Soc. Am.* A **27** 637–47

Zhou G 2009a Fractional Fourier transform of Lorentz–Gaussian beams *J. Opt. Soc. Am.* A **26** 350–5

Zhou G 2009b Fractional Fourier transform of Ince–Gaussian beams *J. Opt. Soc. Am.* A **26** 2586–91

Zhou S, Fu Y, Tang X, Hu S, Chen W and Yang Y 2008 Fourier-based analysis of moiré fringe patterns of superposed gratings in alignment of nanolithography *Opt. Express* **16** 7869–80

IOP Publishing

Principles and Applications of Fourier Optics

Robert K Tyson

Chapter 7

Practical applications

7.1 Pattern recognition and encoding

Taking the concept of a matched filter being able to create a cross-correlation, one can build an optical configuration that will detect whether an optical signal matches a particular optical signal. In one way this can be used to extract specific information from a very noisy background or simply to see if a particular pattern is present in an unknown image.

If a particular signal $s(x, y)$ is sought, then we create an impulse response $h(x, y)$ equal to the inverted complex conjugate of s.

$$h(x, y) = s^*(-x, -y). \tag{7.1}$$

The Fourier transform of this equation is

$$H(f_x, f_y) = S^*(f_x, f_y). \tag{7.2}$$

Based upon the results of the previous chapter, an unknown input $g(x, y)$ can be applied to the impulse response and the result becomes

$$\begin{aligned} v(x, y) &= \iint_{-\infty}^{\infty} h(x - \xi, y - \eta) g(\xi, \eta) \mathrm{d}\xi \, \mathrm{d}\eta \\ &= \iint_{-\infty}^{\infty} g(\xi, \eta) s^*(\xi - x, \eta - y) \mathrm{d}\xi \, \mathrm{d}\eta \end{aligned} \tag{7.3}$$

which is the crosscorrelation of $g(x, y)$ and $s(x, y)$. In the optical configuration of figure 7.1, the frequency plane filter is S^*. If the input signal is exactly s, which is matched to the filter, the field distribution at the filter is exactly $S(f_x, f_y)$. The distribution after the filter is the real function $S(f_x, f_y)S^*(f_x, f_y)$ which is a plane wave. When this is imaged by the final lens one observes a field with a very bright focus.

doi:10.1088/978-0-750-31056-7ch7

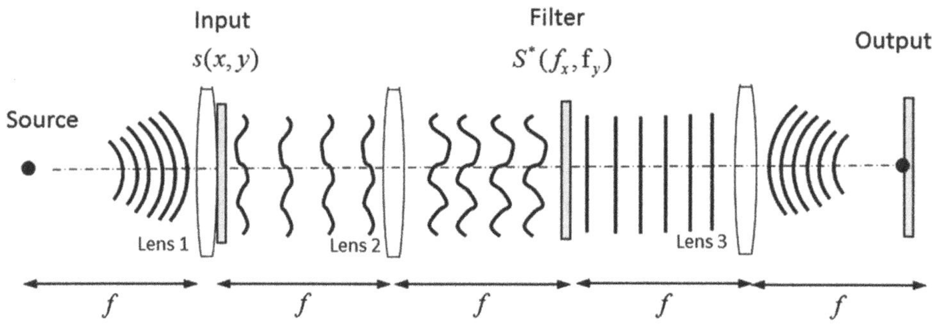

Figure 7.1. Matched filtering operation.

If the input field is not perfectly matched, the output wave has some curvature and the final image is not focused as well. Thus the intensity of the image is an indicator of the presence of the matched signal or a part of it.

One well-known difficulty of the straightforward matched filter approach is the inability to easily detect the signal if it is magnified or rotated. Where Φ is an additional phase factor and c_x and c_y are constants, the impulse response is $H(f_x, f_y) \rightarrow |H(c_x f_x, c_y f_y)| \exp(-i2\pi\, \Phi)$.

Over the decades, dozens of papers have been written about methods to overcome the practical difficulties of the sensitivity to magnification and rotation. Recognizing the similarities between Mellin transforms and Fourier transforms, techniques such as Mellin correlators have been employed (Bryngdahl 1974, Casasent and Psaltis 1977) as well as circular harmonics (Hsu and Arsenault 1982).

Optical character recognition (OCR) has evolved into a variety of groups. OCR targets typewritten text, one glyph or character at a time, or it can target typewritten text, one word at a time, especially for languages that use a space as a word divider. Intelligent character recognition (ICR) targets handwritten print script or cursive text one glyph or character at a time and intelligent word recognition (IWR) also targets handwritten print script or cursive text, one word at a time. This is especially useful for languages where glyphs are not separated in cursive script (Plamondon and Srihari 2000, Arica and Yarman-Vural 2001, Rice *et al* 1999, Cheriet *et al* 2007).

7.2 Image hiding and image encryption

While the previous section addressed finding particular patterns in an image, this section deals with hiding an image within heavy noise or within another image. This 'covered or protected writing' takes on the name derived from the Greek: *steganography*. Closely related to cryptography, steganography is concerned with concealing the fact that a secret message is being sent, as well as concealing the contents of the message; cryptography is the practice of protecting the contents of a message alone. Thus sending a message that appears to be optical, lingual or alphabetical gibberish represents cryptography, sending a message imbedded in the picture of the Eiffel Tower (or a picture of the latest rock star) is steganography.

Originally mentioned over 600 years ago in a book on magic (Trithemius 1499), optical image encryption has come a long way (Peticolas *et al* 1999). Many examples of hidden images can be found in the literature (Wayner 2009). The jargon related to image hiding is straightforward. The *payload* is the data to be covertly communicated. The *carrier* is the signal, stream or data file into which the payload is hidden; which differs from the '*channel*' (typically used to refer to the type of input, such as a JPEG image). The resulting signal, stream or data file which has the payload encoded into it is sometimes referred to as the *package*, *stego file* or *covert message*. The percentage of bytes, samples, or other signal elements which are modified to encode the payload is referred to as the *encoding density* and is typically expressed as a number between 0 and 1. In a set of files, those files considered likely to contain a payload are called *suspects*. If the *suspect* was identified through some type of statistical analysis, it might be referred to as a *candidate*.

The larger the cover message is, in number of bits, relative to the hidden message, the easier it is to hide the latter. For this reason, digital pictures (which contain large amounts of data) are used to hide messages on the internet and on other communication media. It is not clear how commonly this is actually done. For example: a 24 bit bitmap will have 8 bits representing each of the three color values (red, green, and blue) at each pixel. If we consider just the blue there will be 2^8 different values of blue. The difference between 11111111 and 11111110 in the value for blue intensity is likely to be undetectable by the human eye. Therefore, the least significant bit can be used (more or less undetectably) for something else other than color information. If we do it with the green and the red as well we can get one letter of ASCII text for every three pixels.

Stated somewhat more formally, the objective for making steganographic encoding difficult to detect is to ensure that the changes to the carrier (the original signal) due to the injection of the payload (the signal to covertly embed) are visually (and ideally, statistically) negligible; that is to say, the changes are indistinguishable from the noise floor of the carrier. Any medium can be a carrier, but media with a large amount of redundant or compressible information are better suited (Wayner 2009).

From an information theoretical point of view, this means that the channel must have more capacity than the 'surface' signal requires; that is, there must be redundancy. For a digital image, this may be noise from the imaging element. In general, electronics that digitize an analog signal suffer from several noise sources such as thermal noise, flicker noise and shot noise. This noise provides enough variation in the captured digital information that it can be exploited as a noise cover for hidden data. In addition, lossy compression schemes (such as JPEG) always introduce some error into the decompressed data; it is possible to exploit this for steganographic use as well (Mazurczyk and Szczypiorski 2008).

Over the past few decades, optical encryption has been shown to be successful in the spatial domain and the Fourier domain. Double random-phase encoding in the Fresnel domain (Situ and Zhang 2004, Shi *et al* 2006, 2007), in the fractional Fourier domain (Unnikrishnan *et al* 2000, Kong 2013, Liansheng *et al* 2013) and in the discrete cosine transform domain (Chang and Tsan 2005, Zhang *et al* 2007) have all been studied. The use of digital holography in the Fourier domain has been successful (Takai and Mifune 2002, Kishk and Javidi 2003).

7.2.1 Image hiding with a joint transform correlator and hologram

Shi and Zhao propose an optical system whereby the secret information with random phase and key code is encoded into the Fourier hologram of the host image (Shi and Zhao 2011). The joint power spectrum is then used to store or transmit the encoded image. The secret information is encoded in the Fourier domain and can resist various attacks.

Mathematically, it is fairly straightforward. Using the notation of Shi and Zhao (2011), we begin with the host image $g(x, y)$ and the secret image $f(x, y)$. The host is divided into two parts $g_h(x, y)$ and $g_r(x, y)$, both real valued. See figure 7.2. The Fourier hologram is recorded using the optical Fourier transform of $g_h(x, y)$ denoted by

$$G(f_x, f_y) = \iint_{-\infty}^{\infty} g_h(x, y) \exp[-\mathrm{j}2\pi (f_x x + f_y y)] \,\mathrm{d}x\, \mathrm{d}y \tag{7.4}$$

interfered with the reference wave R and inclined by normalized angle a,

$$R(f_x, f_y) = R_0 \exp(\mathrm{j}2\pi a f_x), \tag{7.5}$$

resulting in the intensity of the hologram

$$\begin{aligned} H_0(f_x, f_y) &= |G(f_x, f_y) + R(f_x, f_y)|^2 \\ &= |G(f_x, f_y)|^2 + |R(f_x, f_y)|^2 + G(f_x, f_y) R^*(f_x, f_y) \\ &\quad + G^*(f_x, f_y) R(f_x, f_y). \end{aligned} \tag{7.6}$$

Because the correlation terms are useful, the constant terms are removed.

$$H_0(f_x, f_y) = G(f_x, f_y) R^*(f_x, f_y) + G^*(f_x, f_y) R(f_x, f_y) \tag{7.7}$$

The hidden image $f(x, y)$ is modified by a random-phase code $\alpha(x, y)$ and a key code $b(x, y)$ such that the output to be added to the hologram is

$$f_0(x, y) = \alpha(x, y) f(x, y) + b(x + a, y) \tag{7.8}$$

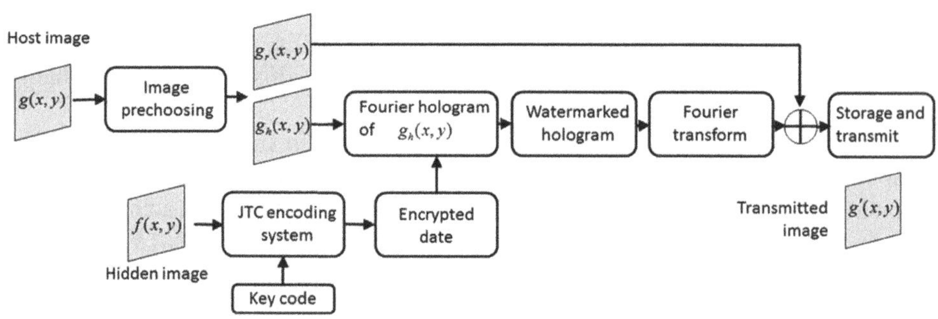

Figure 7.2. Block diagram of image embedding.

whose Fourier transform is

$$F_0(f_x, f_y) = A(f_x, f_y) * F(f_x, f_y) + B(f_x, f_y)\exp[j2\pi(af_x + y)]. \quad (7.9)$$

The result is the joint power spectrum $E_0(f_x, f_y)$ of the hidden image. In the Fourier plane

$$\begin{aligned} E_0(f_x, f_y) = {} & |A(f_x, f_y) * F(f_x, f_y)|^2 + 1 \\ & + |A(f_x, f_y) * F(f_x, f_y)|^* B(f_x, f_y)\exp[j2\pi(af_x + y)] \\ & + |A(f_x, f_y) * F(f_x, f_y)|\, B^*(f_x, f_y)\exp[-j2\pi(af_x + y)]. \end{aligned} \quad (7.10)$$

The inverse Fourier transform of this has four terms, the first two are noise terms centered at ($x = 0$, $y = 0$). They are spatially separated out. Using only the last two terms which are spatially separated out

$$\begin{aligned} E(f_x, f_y) = {} & |A(f_x, f_y) * F(f_x, f_y)|^* B(f_x, f_y)\exp[j2\pi(af_x + y)] \\ & + |A(f_x, f_y) * F(f_x, f_y)|\, B^*(f_x, f_y)\exp[-j2\pi(af_x + y)] \end{aligned} \quad (7.11)$$

and the Fourier hologram embedded with the hidden encoded image, $H_w(f_x, f_y)$ is

$$\begin{aligned} H_w(f_x, f_y) = {} & G(f_x, f_y)R^*(f_x, f_y) + G^*(f_x, f_y)R(f_x, f_y) \\ & + \hat{a}\, |A(f_x, f_y) * F(f_x, f_y)|^* B(f_x, f_y)\exp[j2\pi(af_x + y)] \\ & + \hat{a}\, |A(f_x, f_y) * F(f_x, f_y)|\, B^*(f_x, f_y)\exp[-j2\pi(af_x + y)]. \end{aligned} \quad (7.12)$$

The constant $\hat{a}$ is added to insure the invisibility of the embedded image.

The recovery process is essentially an inverse transform that isolates the host image at ($x = -a$, y) and ($x = +a$, y). See figure 7.3. After inverse transforming we have an image g_h', which contains the original host image, and the keycode

$$g_h'(x, y) = g_h'(x + a, y) + \alpha \bullet b(x, y) \oplus [a(x, y)f(x, y) * \delta(x + a, y)] \quad (7.13)$$

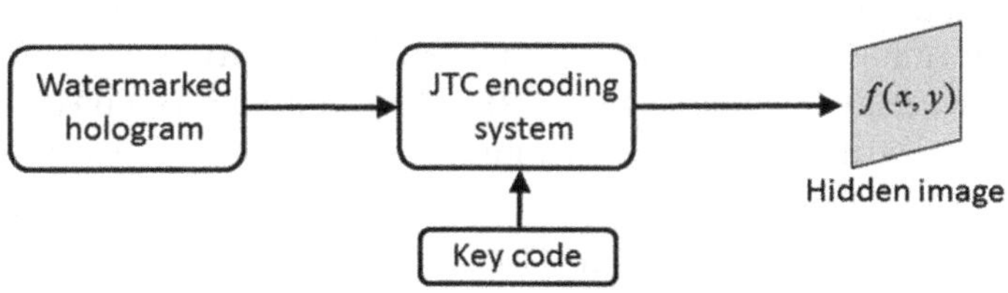

Figure 7.3. Block diagram of hidden image extraction.

where $*$ represents convolution and $\oplus$ represents correlation. The encoded image is placed, or transmitted to, the Fourier plane of the joint transform correlator with the correct keycode and the output plane image is

$$f'(x, y) = \alpha \bullet a(x, y)f(x, y) + g_h(x, y) * b(x, y). \tag{7.14}$$

This image contains the recovered hidden image in the first term and the additive noise in the second term. Because $f(x, y)$ is positive and real, it can be recovered with a CCD device or similar imager. Shi and Zhao (2011) showed the system's validity with computer simulations.

7.2.2 Image hiding with variations of the fractional Fourier transform

Using the basis of a 4-f correlator, each lens along a chain of n lenses acts as a Fourier transforming lens. With $\mathrm{FT}(f(x, y))$ defined as the Fourier transform, the nth iterated Fourier transform is $\mathrm{FT}^n(f) = \mathrm{FT}[\mathrm{FT}^{n-1}(f)]\cdots$. Where $\mathrm{FT}^{-n}(f) = [\mathrm{FT}^{-1}(f)]^n$, the series is finite when n is a non-negative integer. In that case $\mathrm{FT}^0(f) = f$ and the process repeats in the four-period automorphism mentioned in the previous chapter, i.e. $\mathrm{FT}^4(f) = f$.

Using this process optically, a number of images can be encrypted at the various intermediary planes and the encrypted host, with multiple keys, is cascaded and then printed or transmitted (Kong *et al* 2013). The decryption process is roughly the reverse. Liansheng *et al* (2013) show the optical implementation of the encryption that produces the output image $g\ \exp(\mathrm{j}\xi_1)$,

$$g\ \exp(\mathrm{j}\xi_1) = \mathrm{FT}^{-\beta_1}\{\mathrm{FT}^{\alpha_2-\beta_2}[\mathrm{FT}^{\alpha_1}(f\ \exp(\mathrm{j}\phi_1))\exp(\mathrm{j}\phi_2)]\exp(-\mathrm{j}\xi_2)\} \tag{7.15}$$

where $\alpha_1, \alpha_2, \beta_1, \beta_2$ are two different groups of fractional orders and $\phi_1, \phi_2, \xi_1, \xi_2$ are phase functions in the interval $(0, 2\pi)$. The optical decryption follows closely.

Many researchers claim that the multichannel fractional Fourier domain encryption with random phase encoding is highly secure with minimal crosstalk between the multiple images (Unnikrishnan *et al* 2000, Liu *et al* 2001, Jin and Yan 2007, Lin and Shen 2012).

7.3 Image processing and restoration

7.3.1 Image formation

Generalizing the discussion from chapter 5, we will assume that image processing and restoration are two-dimensional processes. We can represent the imaging process by transforming an object $o(x, y)$ incoherent intensity distribution through a linear space-invariant imaging system with a PSF (impulse response) $s(x, y)$ to an image intensity distribution $i(x, y)$. The image formation process is a convolution.

$$i(x, y) = \iint_{-\infty}^{\infty} o(\xi, \eta)\, s(x - \xi, y - \eta)\mathrm{d}\xi\, \mathrm{d}\eta \tag{7.16}$$

7.3.2 Deconvolution: restoration of the object

Now consider that we have the image $i(x, y)$ and, with a known PSF $s(x, y)$, we want to find the unknown object or an estimate of the object's intensity distribution. Using our knowledge of Fourier transforms and convolution, we can take the transform of this equation and produce the image spectrum, the object spectrum and the modulation transfer function, the Fourier transform of the PSF. To keep the problem as simple as possible, we assume that the magnification is unity and there is no symmetry inversion of the image or object.

$$\mathrm{FT}[i(x, y)] = I(f_x, f_y) = \mathrm{FT}[s(x, y) \otimes o(x, y)] = S(f_x, f_y)O(f_x, f_y) \quad (7.17)$$

Solving for the object spectrum is trivial,

$$O(f_x, f_y) = \frac{I(f_x, f_y)}{S(f_x, f_y)} \quad (7.18)$$

and the estimated object is recovered

$$\hat{o}(x, y) = \mathrm{FT}^{-1}[O(f_x, f_y)]. \quad (7.19)$$

Thus, multiplying the image spectrum by the 'inverse filter'

$$H(f_x, f_y) = \frac{1}{S(f_x, f_y)} \quad (7.20)$$

recovers the object. Two problems make this approach unworkable from a practical standpoint. First, diffraction and aberrations limit the range where the transfer function $S(f_x, f_y)$ is non-zero and the filter becomes undefined. Diffraction *necessarily* blurs the image somewhat by cutting out high spatial frequencies. Aberrations can remove many low frequencies as well. Second, the inverse filter amplifies the frequency components with the lowest signal-to-noise ratios (SNRs) such that the recovered object is dominated by noise in the image. If we now explicitly consider the noise $n(x, y)$, our expression for the imaging process becomes

$$i(x, y) = o(x, y) \otimes s(x, y) + n(x, y) \quad (7.21)$$

where we assume that the noise is an additive random process. Now we try to find a filter that minimizes the difference between the unknown actual object $o(x, y)$ and our estimate of the object $\hat{o}(x, y)$. We further assume, both for the object and for the noise, that we know the distributions of average power over frequency, i.e. the power spectral densities, $\Phi_o(f_x, f_y)$ and $\Phi_n(f_x, f_y)$. Following the derivation of Gonzalez and Woods (1992), the optimum restoration filter, the Wiener filter, is found.

$$H(f_x, f_y) = \frac{S^*(f_x, f_y)}{|S(f_x, f_y)|^2 + \dfrac{\Phi_n(f_x, f_y)}{\Phi_o(f_x, f_y)}} \quad (7.22)$$

The ratio of power spectral densities (PSDs) in the denominator corresponds to the inverse of the SNR. When the SNR is high, the Wiener filter reduces to the simple inverse filter above.

$$H = \frac{S^*}{|S|^2} = \frac{1}{S} \tag{7.23}$$

7.3.3 Using *a priori* information

It has been shown recently that there are some algorithms that can overcome some of the difficulties of low values of $S(f_x, f_y)$. Flaherty (2009) notes that missing information, in terms of very low values for $S(f_x, f_y)$, can be recovered if it is known that the physically measurable parameters, such as light intensities, are positive definite. He shows that the object consists of the inverse of the known filtered object spectrum plus the inverse of the object spectrum at frequencies where the transfer function is undefined. The object is the sum

$$o(x, y) = \hat{o}(x, y) + u(x, y) \tag{7.24}$$

where the function $u(x, y)$ is the inverse transform of the unknown coefficients of the object spectrum.

$$u(x, y) = \mathrm{FT}^{-1}[U(f_x, f_y)]. \tag{7.25}$$

The known and unknown Fourier coefficients are mutually exclusive, i.e. $\hat{O}(f_x, f_y)U(f_x, f_y) = 0$. The method then employs an iterative process to estimate the object. The first approximation is found by taking the Fourier transform of the positive part of $\hat{o}(x, y)$ and then inverse transforming to get an estimate for $o(x, y)$. The Fourier transform of this $o(x, y)$ is then calculated. The $O(f_x, f_y)$ coefficients are replaced by the values from the original data and the new $U(f_x, f_y)$ coefficients are retained for the inverse Fourier transform. This process is repeated until the coefficients converge to the desired tolerance.

7.3.4 Spatially filtering images

The two most common image enhancement procedures are smoothing and edge enhancement. They are easily carried out by digitally or optically filtering the images to create an enhanced image. Where the spatial filter is $S(f_x, f_y)$, the original image appears as the object in the convolution process. The enhanced image is the spatially filtered new image.

$$i(x, y) = \mathrm{FT}^{-1}[I(f_x, f_y)] = \mathrm{FT}^{-1}[S(f_x, f_y)O(f_x, f_y)]. \tag{7.26}$$

Smoothing is a natural result of diffraction in that high spatial frequencies are not transferred to the image. While this blurring process can be the limiting factor in image resolution, it also can be intentionally added to get rid of high frequency noise such as speckle or other image artifacts like dead pixels, etc. The following figures show the result of low-pass filtering an image. The sharp edges of figure 7.4, which contain very high frequency information, are smoothed by the low-pass filter. The result is figure 7.5. The absolute magnitude of the Fourier transform of the object is shown in figure 7.6. A circularly symmetric filter with a hard cutoff frequency is represented by

$$\begin{aligned} S(f_x, f_y) &= 1 \qquad \sqrt{f_x^2 + f_y^2} \leqslant f_{\text{cutoff}} \\ S(f_x, f_y) &= 0 \qquad \sqrt{f_x^2 + f_y^2} > f_{\text{cutoff}}. \end{aligned} \tag{7.27}$$

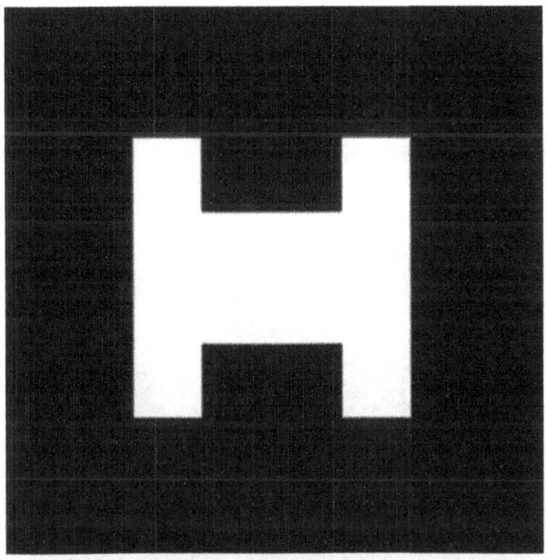

Figure 7.4. Test object.

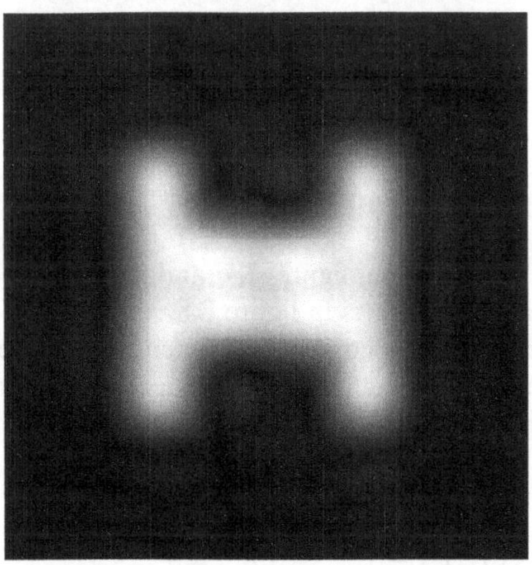

Figure 7.5. Filtered, smoothed representation of the test object.

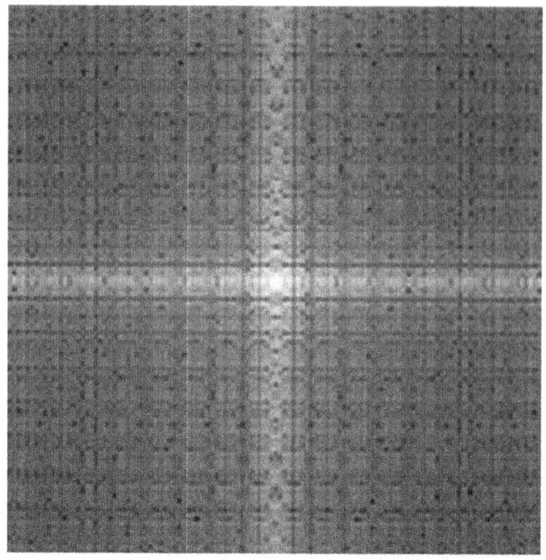

Figure 7.6. Modulus of the Fourier transform of the test object. The logarithm of the modulus is shown for clarity.

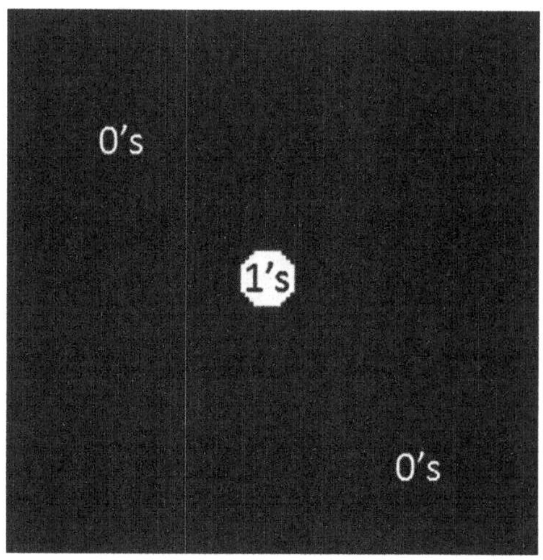

Figure 7.7. Low-pass circular spatial filter.

The spatial filter is shown in figure 7.7 and the absolute magnitude of the transform of the enhanced image is in figure 7.8.

The opposite of smoothing is edge enhancement where the smooth parts of the image are removed and only high spatial frequency components or sharp edges remain. The sharp high-pass filter with a sharp cut-on frequency is

$$\begin{aligned} S(f_x, f_y) &= 0 \qquad \sqrt{f_x^2 + f_y^2} \leqslant f_{\text{cut-on}} \\ S(f_x, f_y) &= 1 \qquad \sqrt{f_x^2 + f_y^2} > f_{\text{cut-on}}. \end{aligned} \tag{7.28}$$

Figure 7.8. Modulus of the Fourier transform of the smoothed test object.

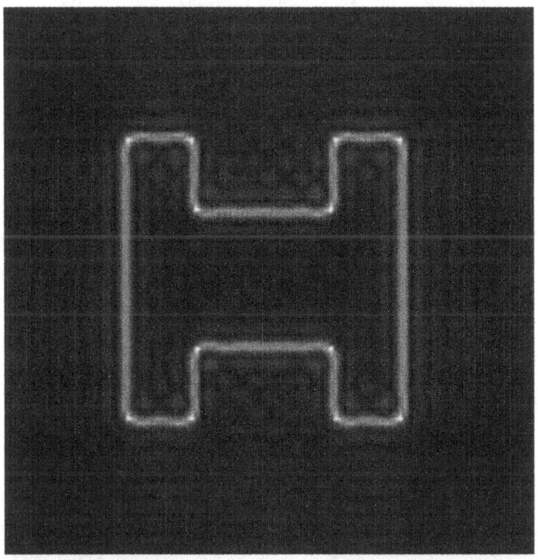

Figure 7.9. Edge-enhanced spatially filtered test object.

The edge enhancement result is shown in figure 7.9. The Fourier transform of the object is the same as figure 7.6; the spatial filter is shown in figure 7.10; and the transform of the enhanced image is in figure 7.11.

One interesting effect that exposes itself with edge enhancement is the appearance of 'age' enhancement. By removing the smooth features of a human face, the wrinkles are enhanced and the subject appears older. Using the same filter as figure 7.10, figures 7.12(*a*) and (*b*) show the effect of edge enhancement on the famous subject, Lena.

While the simple examples shown here are done digitally, they suffer from the fact that large data sets (high resolution images) require large processing times. To overcome this, it has been shown that all-optical processing could essentially make instantaneous edge enhancements, or at least as fast as video rates (Casasent and

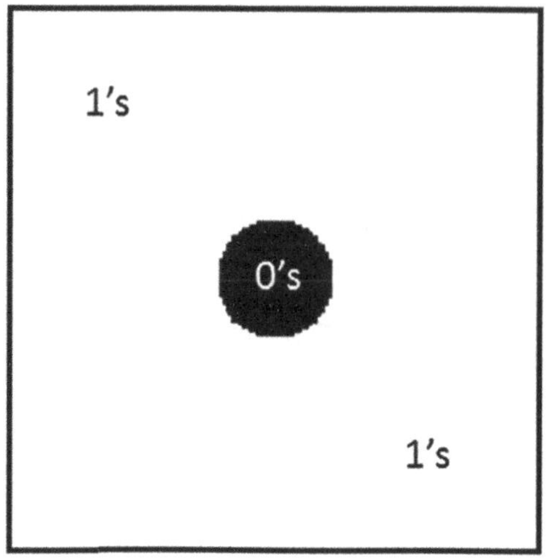

Figure 7.10. High-pass circular spatial filter.

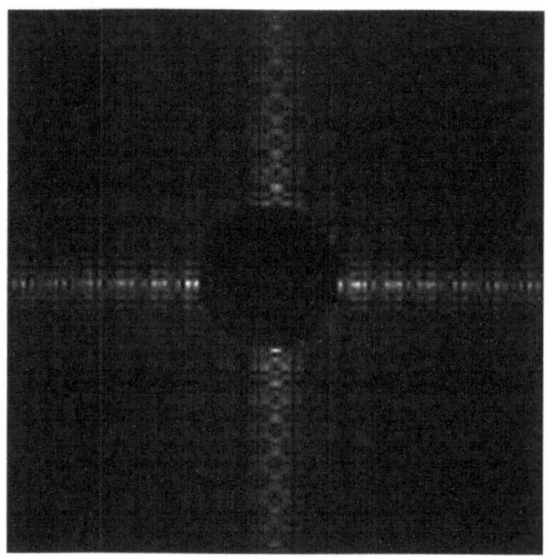

Figure 7.11. Modulus of the Fourier transform of the edge-enhanced test object.

Chen 1983). While a number of all-optical techniques have been proposed (Yelleswarapu *et al* 2008, Mazzaferri and Ledesma 2007, Ferrari *et al* 2010), one recent advance has been shown to be promising (Ferrari and Flores 2010). The technique takes advantage of the polarization properties of twisted-nematic liquid crystal displays, which are manufactured to work between a polarizer analyzer pair. The image will have different contrast for different polarizations. One will be a replica of the original image and the other will be a contrast-reversed (negative) replica. The negative replica is low-pass filtered with a 4-f optical processor and then both images are recombined by a beamsplitter. The result is an image with enhanced second derivatives or equivalently, enhanced edges. See figure 7.13. Another recently

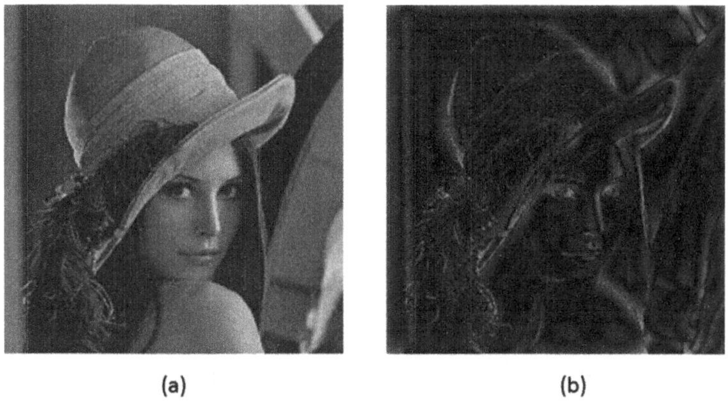

Figure 7.12. (a) Image of 'Lena'. (b) Edge-enhanced image of 'Lena'.

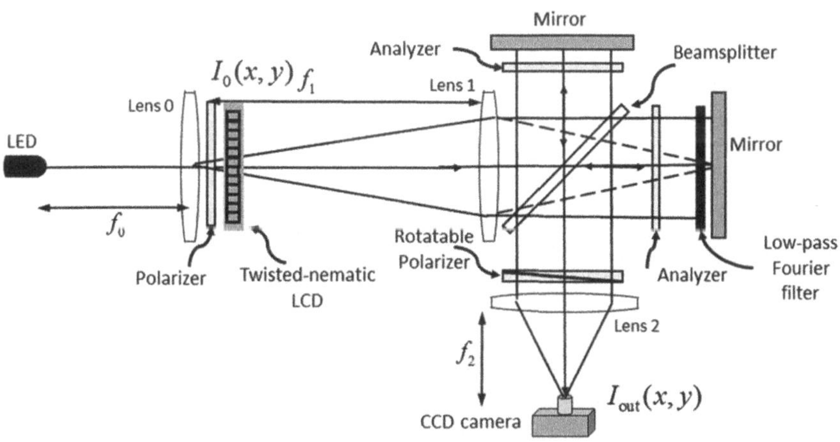

Figure 7.13. Optical layout for edge enhancement with a low-pass Fourier filter.

reported study uses specifically designed *meta*materials to perform 'photonic calculus' such that a light wave passing through the material emerges as its spatial derivative, creating an all-optical edge detection technique (Silva *et al* 2014).

7.3.5 Phase imaging

There are many situations where the object to be imaged does not clearly have any intensity contrast to perform conventional imaging such as photography. In crystallography, diffractive optics manufacturing and especially biology and medicine, there exist countless objects of interest that are transparent, or nearly so. Subtle variations in the index of refraction throughout the object require interferometric or spatial filtering methods. In interferometry, a well-characterized reference wave is interfered with light transmitted through the phase object and an interference pattern is observed. The fringes contain information about the phase object. However, if the phase disturbance through the sample exceeds π radians, the phase-to-intensity map is ambiguous (Teschke *et al* 2008). This difficulty can be overcome in a variety of ways. Classical phase-shifting interferometry can be enhanced by coupling a Mach–Zehnder interferometer with a 4-f arrangement using a phase grating in the Fourier plane (Serrano-Garcia *et al* 2012). While three phase shifts (i.e. four interferograms) are shown experimentally, the authors show how an extension to nine interferograms is possible. Much research exists concerning the phase unwrapping problem. A summary of the first phase of the field can be found in the book by Ghiglia and Pritt (1998).

Spatial filtering methods make use of Fourier theory. Light propagated through the object is treated as both diffracted and undiffracted light. A Fourier transforming lens can spatially separate the diffracted light from the undiffracted light in the Fourier plane where each can be subjected to a filter (Zernike 1942a, 1942b). For a linear phase-to-intensity mapping, the phase contrast must be less than $\frac{\pi}{3}$ (Gluckstad and Mogensen 2001). To extend the limitations on the phase contrast, various new approaches have been employed. One method expands the phase object $O(\phi)$ in a Taylor expansion

$$O(\phi) = \exp(\mathrm{j}\phi) = 1 + \mathrm{j}\phi - \frac{1}{2}\phi^2 - \frac{1}{6}\mathrm{j}\phi^3 + \frac{1}{24}\phi^4 + \cdots \tag{7.29}$$

and carries as many terms in the expansion as necessary, while making use of a variable Fourier filter in the process (Tesche and Sinzinger 2009). Because of the need for imaging biological processes, there has been extensive work on phase unwrapping in recent years (Schofield and Zhu 2003, Arines 2003, Li and Su 2005). One extension of the phase imaging process is dynamic phase imaging. When the objects of interest are smaller than $\frac{\lambda}{2}$, such as viruses or nanometric magnetic domains in magnetic films, UV or x-ray light is required. Both high two-dimensional spatial and one-dimensional temporal resolution are needed. Recent work presented a new algorithm where high temporal resolution can be obtained without high power sources (that can destroy the object or distort the mechanism of interest) if the mechanism is assumed to be slowly varying from frame to frame. In that way, the 'greedy algorithm for sparse phase-retrieval' can be employed (Shechtman *et al* 2013).

In a related technique, when the image from a microscope is transferred to a phase retrieval algorithm in a computer, the patented method is referred to as ptychography. A number of systems that automate this technique are available commercially (Ou *et al* 2013, Zheng *et al* 2013).

Another multidimensional problem, which is three-dimensional spatial resolution of thin phase objects, is addressed with a tomography solution. Conventional back-projection, the inverse Radon transform (Deans 1983) is used with a 4-f optical transform correlator to provide a directional edge enhancement at each slice (Meneses-Fabian *et al* 2011).

7.3.6 Schlieren

Schlieren are techniques to visualize optical inhomogeneities such as refractive index gradients in transparent material not necessarily visible to the human eye. Schlieren is German for 'streak'. It can show the distribution of temperature, concentrations or pressure in air flows, wind tunnels, aircraft shock waves and many other applications (Settles 2001). It can be performed with inexpensive incoherent white light as was first observed by Robert Hooke in 1665 using two candles (Hooke 1665). One candle provided the source and the warm air rising from the second candle produced the schliere. Schlieren can also be observed with sunlight or halogen or xenon lamps. Because of the size of the objects it can visualize, it can be preferable to other methods such as interferometry.

In coherent image processing, a lens produces a Fourier transform of the object. If the object distribution is given by

$$O(x, y) = A(x, y)\exp[j\phi(x, y)], \tag{7.30}$$

when the phase $\phi(x, y)$ is constant, it produces an amplitude image. When the amplitude $A(x, y)$ is constant, we have a phase image, not visible to our eyes or photodetectors. With a suitable optical filter in the Fourier plane, one can convert the phase changes into an amplitude or a phase-dependent intensity. In an optical coherent system, zero-order filtering or strioscopy is performed by applying a black dot at the center of the Fourier transform, thereby removing the zero-order frequency. A half-plane filter cuts one half of the Fourier transform which reveals the schlieren technique. For small phase images

$$O(x, y) = \exp[j\phi(x, y)] \approx 1 + j\phi(x, y), \tag{7.31}$$

the resulting strioscope visualized image is

$$I(x, y) = \phi^2(x, y), \tag{7.32}$$

and the schlieren image is

$$I(x, y) \approx 1 - \frac{2}{\pi x} \otimes \phi(x, y). \tag{7.33}$$

A Fourier optics description of the focusing schlieren technique has been presented previously (Serafino and Sirotti 2002). Building up the incoherent light source as a set of coherent sources, many colored Fourier transforms appear in the conjugate plane of the light source. In the Fourier plane, the convolution of the image of the source and the Fourier transform of the image appears.

The optical system is shown in figure 7.14. The source is coded by using a source grid composed of $2N + 1$ open slits of width $2a$, closed slits of width $2b$ and step size $2a + 2b$. The grid is described by

$$r_S = \text{rect}\left(\frac{x}{a}\right) \otimes \sum_{k=-N}^{N} \delta[x - 2k(a + b)]. \tag{7.34}$$

The cutoff grid placed in the transform plane is the opposite of the source grid,

$$r_C = \text{rect}\left(\frac{x}{a}\right) \otimes \sum_{k=-N}^{N} \delta[x - (2k) + 1(a + b)] \tag{7.35}$$

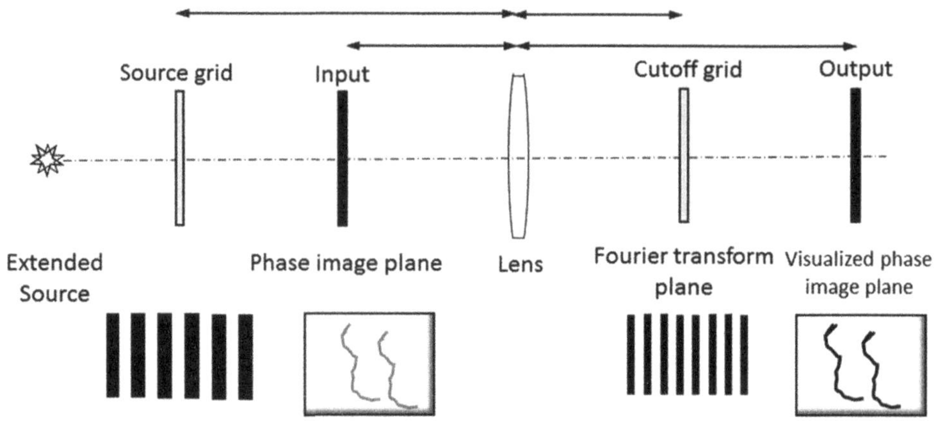

Figure 7.14. Schlieren system. The Fourier plane is the conjugate of the source grid. The visualized phase image is the conjugate of the input.

The Fourier transforms of the grids are

$$R_S(f_x, f_y) = 2a\,\mathrm{sinc}(af_x)\sum_{k=-N}^{N}\exp[-\mathrm{j}(a+b)(2kf_x)] \tag{7.36}$$

and

$$R_C(f_x, f_y) = 2b\,\mathrm{sinc}(bf_x)\sum_{k=-N}^{N}\exp[-\mathrm{j}(a+b)(2k+1)f_x]. \tag{7.37}$$

The corresponding light intensity distributions produced by the grids are

$$\begin{aligned} I_S(f_x, f_y) &= R_S^*(f_x, f_y)R_S(f_x, f_y) = 4a^2\mathrm{sinc}^2(af_x)C^2(f_x) \\ I_C(f_x, f_y) &= R_C^*(f_x, f_y)R_C(f_x, f_y) = 4b^2\mathrm{sinc}^2(bf_x)C^2(f_x) \end{aligned} \tag{7.38}$$

where

$$C(f_x) = 2\frac{\sin(a+b)(N+1)f_x}{\sin(a+b)f_x}\cos(a+b)Nf_x - 1. \tag{7.39}$$

If the cutoff grid is completely superimposed on the image of the source grid, with incoherent quasi-monochromatic sources and small refractive phase objects, the grid operates as a high-pass filter in the x-direction. If the cutoff grid is partially shifted on the image of the source grid, three terms appear. The first is the light coming from the background, the second term is caused by the symmetric portions of the Fourier transforms without low frequencies and the third is due to the nonsymmetric cutoff of the Fourier transforms. If the offset is small, the first and third components are negligible. If the offset is large, the second term is repressed and the first and third terms dominate and a situation similar to strioscopy appears, thereby demonstrating how schlieren images can be obtained with white light but analyzed as a coherent process (Serafino and Sirotti 2000).

7.4 Holography

7.4.1 Introduction

Holography is a technique whereby the simultaneous recording of the amplitude and the phase of light diffracted or scattered from an object can be reconstructed maintaining the three-dimensional nature of the object. Upon reconstruction, an observer can change their viewing orientation and the image appears to be a multi-dimensional solid.

The holographic recording itself is not an image but rather an interference pattern between the light from the object and a reference wave, typically a laser. Holography (from the Greek words *hólos* and *grafē* meaning 'whole writing') was invented by physicist Dennis Gabor in 1948 (Gabor 1948). Built on the pioneering work in x-ray microscopy, holography, arguably the most important application of Fourier optics, was recognized by the Royal Swedish Academy of Sciences who awarded the Nobel Prize in Physics to Gabor in 1971.

Many volumes have been written on the history and development of holography, such as the classic text by Collier *et al* (1971) and a book on practical engineering of holograms (Saxby 2012). It is not necessary to repeat the extent of the material here, but rather just introduce the principles and discuss some modern applications.

7.4.2 Types of holograms

Several types of holograms can be made. Transmission holograms, such as those produced by Leith and Upatnieks (1962), are viewed by shining laser light through them and looking at the reconstructed image from the side of the hologram opposite the source. A later refinement, the 'rainbow transmission' hologram, allows more convenient illumination by white light rather than by lasers. Rainbow holograms are commonly used for security and authentication, for example, on credit cards and product packaging.

7.4.3 Mathematical principles

Using an optical configuration as shown in figure 7.15, a recording on a photographic plate or other intensity-sensitive recording medium is made into a

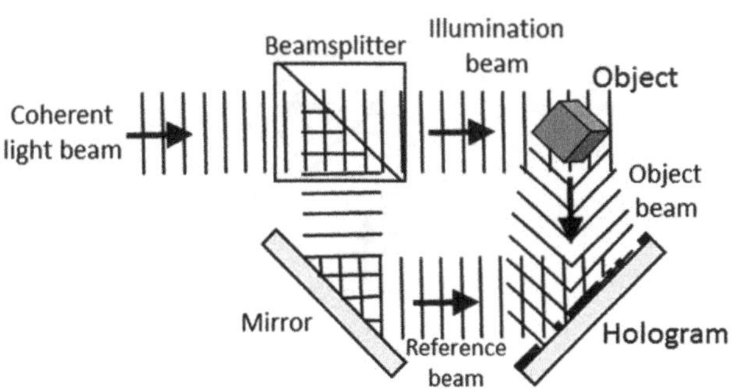

Figure 7.15. Recording a hologram.

hologram. If the field reflected from the object is U_{Obj} represented by its amplitude and phase

$$U_{\mathrm{Obj}}(x, y) = |A_{\mathrm{Obj}}(x, y)| \exp[\mathrm{j}\phi_{\mathrm{Obj}}(x, y)] \tag{7.40}$$

and a reference beam with which it interferes is

$$U_{\mathrm{ref}}(x, y) = |A_{\mathrm{ref}}(x, y)| \exp[\mathrm{j}\phi_{\mathrm{ref}}(x, y)], \tag{7.41}$$

the intensity of the sum of the two waves is

$$\begin{aligned} I(x, y) = {} & |A_{\mathrm{ref}}(x, y)|^2 + |A_{\mathrm{Obj}}(x, y)|^2 \\ & + 2\,|A_{\mathrm{ref}}(x, y)||A_{\mathrm{Obj}}(x, y)|\cos[\phi_{\mathrm{ref}}(x, y) - \phi_{\mathrm{Obj}}(x, y)]. \end{aligned} \tag{7.42}$$

The hologram that is recorded takes into account the ratio between the response of the recording medium and the incident intensity β' and the bias t_{bias} of the constant reference exposure. With a uniform reference intensity $|A_{\mathrm{ref}}|^2$, the recorded hologram then becomes

$$t_{\mathrm{holo}}(x, y) = t_{\mathrm{bias}} + \beta'(|A_{\mathrm{Obj}}|^2 + A_{\mathrm{ref}}^{*} A_{\mathrm{Obj}} + A_{\mathrm{ref}} A_{\mathrm{Obj}}^{*}). \tag{7.43}$$

Now, using the configuration of figure 7.16, the original wavefront can be reconstructed and an image of the original object, including phase (depth or viewing angle) information, is observed.

$$B(x, y)t_{\mathrm{holo}}(x, y) = Bt_{\mathrm{bias}} + \beta' B\,|A_{\mathrm{Obj}}|^2 + \beta' BA_{\mathrm{ref}}^{*} A_{\mathrm{Obj}} + \beta' BA_{\mathrm{ref}} A_{\mathrm{Obj}}^{*} \tag{7.44}$$

If the reconstruction wave is an exact replica of the reference wave $B = U_{\mathrm{ref}}$, the third term becomes

$$U_3(x, y) = \beta' |A_{\mathrm{ref}} A_{\mathrm{ref}}^{*}|\ A_{\mathrm{Obj}}(x, y) = \beta' |A_{\mathrm{ref}}|^2 A_{\mathrm{Obj}}(x, y) \tag{7.45}$$

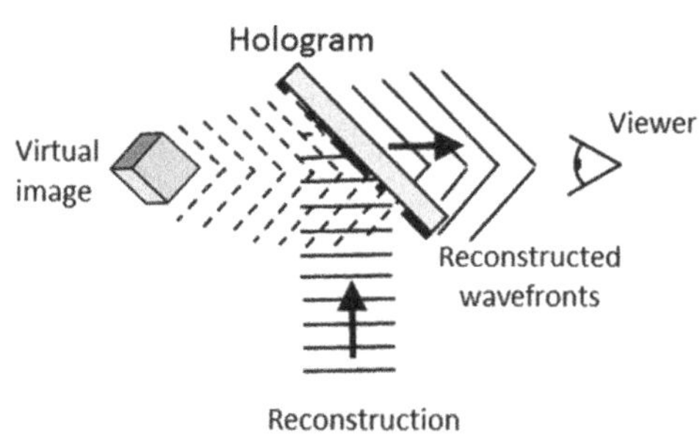

Figure 7.16. Reconstructing the image from the hologram.

which, up to the multiplicative constant, is an exact duplication of the original object wavefront. If the reconstruction wave is the complex conjugate of the reference wavefront, the fourth term becomes

$$U_4(x, y) = \beta' |A_{\text{ref}} A^*_{\text{ref}}| \, A^*_{\text{Obj}}(x, y) = \beta' |A_{\text{ref}}|^2 A^*_{\text{Obj}}(x, y). \tag{7.46}$$

Depending upon whether the reconstruction beam is exact or a conjugate of the reference wave, the image will appear to be virtual behind the hologram from the observer or real in front of the hologram on the side of the observer. In either case the images are accompanied by a coherent background Bt_{bias}.

7.4.4 Applications

Over the years since Gabor's original recorded hologram and the development of the Leith–Upatnieks offset-reference hologram (Leith and Upatnieks 1962), thousands of refinements, advancements and applications have been seen.

Like the example above, when the light recorded is in the image plane, it is called an *image hologram.* Other amplitude holograms can be simply classified. If the light that is recorded is in the Fresnel diffraction region of the object, it is a *Fresnel hologram.* Similarly, if the light that is recorded is in the Fraunhofer diffraction region of the object, it is a *Fraunhofer hologram.* And finally, when the light that is recorded is in the Fourier plane of the object, it is a *Fourier hologram.*

If the material in the object is transparent with a varying index of refraction, a purely phase hologram can be recorded, which is an efficient way to generate phase gratings. A thin hologram is one where the thickness of the recording medium is much less than the spacing of the interference fringes which make up the holographic recording. A thick or volume hologram is one where the thickness of the recording medium is greater than the spacing of the interference pattern. The recorded hologram is now a three-dimensional structure, and it can be shown that incident light is diffracted by the grating only at a particular angle, the Bragg angle.

7.4.5 Further applications of holography

Microscopy and high-resolution volume imagery. Three-dimensional visualization of microscopic phase objects (Tishko *et al* 2005, 2010).

Holographic interferometry. Holographic interferometry is a technique that enables static and dynamic displacements of objects with optically rough surfaces to be measured to optical interferometric precision (i.e. to fractions of a wavelength of light) (Jones and Wykes 1989). It can also be used to detect optical path length variations in transparent media, which enables, for example, fluid flow to be visualized and analyzed. It can also be used to generate contours representing the form of the surface. It has been widely used to measure stress, strain and vibration in engineering structures.

Holographic optical elements (HOEs). HOEs involve using holography to create optical elements such as gratings, beamsplitters, diffuse reflectors and other conventional optical elements (Jones and Wykes 1989).

Real-time interferometry and vibration analysis. This is the study of the interference fringes generated when a hologram is made of an object and is later placed back into its original position relative to the object, now slightly deformed, so that there is interference between the object and its hologram (Nakano and Minami 1985, Leith and Upatnieks 1970).

Imaging through distorting media. This technique involves using the characteristics of nonlinear materials to create wavefront reconstruction after passage of light through a distorting or aberrating medium (Ransom 1980). A particularly promising application is optical phase conjugation. It allows the removal of the wavefront distortions a light beam receives when passing through an aberrating medium, by sending it back through the same aberrating medium with a conjugated phase (Yariv and Pepper 1977). Nonlinear optical four-wave mixing is a type of real-time holography.

Dynamic imaging through distorting media (adaptive optics). Linear optical phase conjugation (adaptive optics) is a near real-time method (Tyson 2011a, 2011b). This is useful, for example, in free-space optical (FSO) communications to compensate for atmospheric turbulence (the phenomenon that gives rise to the twinkling of starlight) (Tyson 1996) or in compensation for the turbulence over an aero-optical window (Goorskey 2013).

Dynamic imaging. There also exist holographic materials that do not need the developing process and can record a hologram in a very short time (Teng *et al* 2014). This allows one to use holography to perform some simple operations in an all-optical way (Plesniak *et al* 2001).

Examples of applications of such real-time holograms include phase-conjugate mirrors (time-reversal of light) (Pepper 1986), optical cache memories (Esener 1989, De Caro *et al* 1991), image processing (pattern recognition of time-varying images) (Cappellini 1997) and optical computing (Hillerkuss *et al* 2011).

Holographic data storage. Holography can be put to a variety of uses other than recording images. Holographic data storage (Ogai *et al* 1993) is a technique that can store information at high density inside crystals or photopolymers. The ability to store large amounts of information in some kind of media is of great importance, as many electronic products incorporate storage devices (Heanue *et al* 1994). As current storage techniques such as Blu-ray Discs reach the limit of possible data density (due to the diffraction-limited size of the writing beams), holographic storage has the potential to become the next generation of popular storage media (Knight 2005). Holographic scanners are in use in post offices, shipping firms and automated conveyor systems to determine the three-dimensional size of a package.

Holographic data security. Security holograms are very difficult to forge because they are replicated from a master hologram that requires expensive, specialized and technologically advanced equipment. They are used widely in many currencies and in credit and bank cards as well as passports, ID cards, books, DVDs and sports equipment (Matoba and Javidi 2001, Unnikrishnan *et al* 1998).

Digital holography. This is the technology of acquiring and processing holographic measurement data, typically via a CCD camera or a similar device. In particular, this includes the numerical reconstruction of object *data* from the recorded measurement data, in distinction to an optical reconstruction which reproduces an *aspect* of the object. Digital holography typically delivers three-dimensional surface or optical thickness data (Schnars and Jüptner 2005).

Holographic neural networks. This involves creating neural networks with a single hologram and application to non-iterative computation (Li *et al* 1993, Wilson *et al* 1997).

Holographic displays and holographic art. This is three-dimensional dynamic volume holography (e.g. Princess Leia in *Star Wars*) (Yaraş *et al* 2010, Connor 2014).

7.5 Fourier optics in optical communications

Optical communications have been around for millennia. From basic sign language to smoke signals to semaphores to modern fiber optical networks, the transfer of information on a beam of light is central to civilization. Both FSO propagation and guided-wave systems are important and the differences are significant. The natural modes of FSO signals are the Fourier components of plane waves travelling in different directions. In a bounded dielectric medium such as a fiber, the modes are determined by the geometry and refractive index of the materials.

7.5.1 Fiber optics transmission

A section of optical fiber is shown in figure 7.17. The radius of the core of material of refractive index n_1 is a and the outer radius of the cladding of index n_2 is b. If the external medium is air, and the maximum (half) angle from the fiber axis is θ_a, the numerical aperture (the angular spread of the light entering and exiting the fiber) is

$$NA_{\text{air}} = \sin\theta_a = \sqrt{n_1^2 - n_2^2}. \tag{7.47}$$

If the fractional difference of refractive index between the core and the cladding is $\Delta = (n_1 - n_2)/n_1$, the numerical aperture is approximately

$$NA_{\text{air}} \approx n_1\sqrt{2\Delta}. \tag{7.48}$$

For the core, the numerical aperture is

$$NA_{\text{core}} = \sqrt{\frac{n_1^2 - n_2^2}{n_1^2}} \approx \sqrt{2\Delta}. \tag{7.49}$$

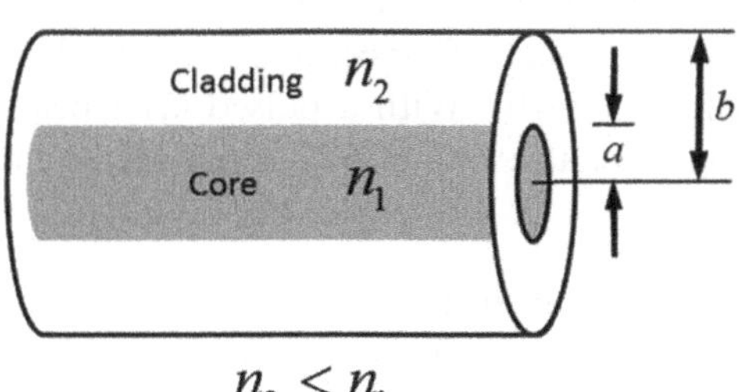

Figure 7.17. Geometry of an optical fiber.

Because the index of refraction typically depends upon wavelength, dispersion in the fiber degrades the pulses of light which carry the information. If the field of the original signal $U(t)$ begins with a complex amplitude and phase, the signal in the fiber $u(t)$ becomes

$$u(t) = U(t)[-\mathrm{j}(\omega t - \beta(\omega)L]) \tag{7.50}$$

where $\omega = 2\pi\nu$ is the angular optical frequency, L is the length of the fiber and $\beta(\omega)$ is the frequency dependent propagation constant. For a pulse of spectral width $\Delta\omega$, the group velocity of the pulse is $v_g = \frac{\partial\omega}{\partial\beta}$ and the time-spreading of the pulse $\Delta\tau$ is

$$\Delta\tau = \frac{\partial^2\beta}{\partial\omega^2}L\ \Delta\omega. \tag{7.51}$$

Details of the methods to compensate or overcome dispersion, coupling light from a fiber or between fibers, and other engineering strategies can be found in the literature (Crisp 2005, Agrawal 2010, Hayes 2006).

7.5.2 Free-space optical communications

FSO technology sends information on light propagating in free space, which includes vacuum, air or other transmitting gases (Bouchet *et al* 2006).

7.5.2.1 Short-range links

Free-space point-to-point optical links can be implemented using visible or infrared laser light, although low-data-rate communication over short distances is possible using **LEDs** (Aschenbrenner 2010, Grubor *et al* 2008). In some cases very short distances need to be crossed, such as interchip communication channels. Optical methods are especially useful (Xue *et al* 2010).

7.5.2.2 Vacuum links

In outer space, the communication range of FSO communication is currently of the order of several thousand kilometers, but has the potential to bridge interplanetary distances of millions of kilometers, using optical telescopes as beam expanders. In January 2013, NASA used lasers to beam an image of the Mona Lisa to the Lunar Reconnaissance Orbiter roughly 240 000 miles away (Peckham 2013).

The distance records for optical communications involve detection and emission of laser light by space probes. The Lunar Laser Communications Demonstration (LLCD), a technology flying aboard NASA's Lunar Atmosphere and Dust Environment Explorer (LADEE) spacecraft, recently set a record for the fastest download rate between the Moon and Earth. With a pulsed laser beam, the instrument sent data over 384 000 kilometers at a rate of 622 Mbps (Kramer 2013).

7.5.2.3 Terrestrial links

One principal advantage of optical links is security. Laser beams are very narrow and can make point-to-point propagation without the large angular spread of radio or microwaves. In addition to conventional encoding, secure FSO communications have been proposed using a laser N-slit interferometer where the laser signal takes the form of an interferometric pattern. Any attempt to intercept the signal causes the collapse of the interferometric pattern (Duarte 2005). This technique has been demonstrated to work over propagation distances of practical interest and, in principle, it could be applied over large distances in space (Duarte *et al* 2011).

The maximum range for terrestrial links is of the order of a few kilometers and is limited by absorption, dispersion, turbulence-induced scintillation, background interference (sunlight), aerosols, fog and precipitation. These factors cause an attenuated receiver signal and lead to a higher bit error rate (BER). To compensate for atmospheric interference, error correction algorithms, similar to those used in CDs, were implemented (Anguita 2005). To overcome the other atmosphere-related issues, some solutions such as multi-beam or multi-path architectures use more than one sender and more than one receiver. Some state-of-the-art devices also have larger fade margin (extra power, reserved for rain, smog, fog). A comprehensive analysis of data was used to determine the transfer function of an atmospheric channel (Reinhardt *et al* 2012).

The BER increase occurs when a beam is directed from the transmitter to the receiver along a path with optical atmospheric turbulence. Because of the varying indices of refraction along the various paths, the signal (s) sees destructive interference (fades) at times and constructive interference (surges) at other times. When the intensity of the expected signal is $\langle i_s \rangle$, the probability distribution of this interference process $p_I(s)$ is a gamma–gamma distribution (Al-Habash *et al* 2001).

$$p_I(s) = \frac{2(\alpha\beta)^{(\alpha+\beta)2^{-1}}}{\Gamma(\alpha)\Gamma(\beta)\langle i_s \rangle}\left(\frac{s}{\langle i_s \rangle}\right)^{[(\alpha+\beta)2^{-1}]-1} K_{\alpha-\beta}\left(2\sqrt{\frac{\alpha\beta s}{\langle i_s \rangle}}\right) \tag{7.52}$$

The gamma functions $\Gamma()$ and the modified Bessel function of the second kind $K_n()$ are described elsewhere (Abramowtz and Stegun 1972) and α and β are determined by atmospheric turbulence strength and propagation scenarios (Andrews *et al* 2001). This distribution ranges from a zero (off) signal to signals a few times the expected signal $\langle i_s \rangle$. The difficulty arises when the noise from the detector overlaps the signal distribution in the low end and false negatives or false positive result in bit error. The probability of the noise assumes a Gaussian distribution

$$P(s, \text{SNR}) = \frac{1}{2}\,\text{erfc}\left(\frac{i_s}{2\sqrt{2}\,\sigma_{\text{Noise}}}\right). \tag{7.53}$$

The complementary error function erfc( ) is defined elsewhere (Abramowitz and Stegun 1972).

The BER for amplitude modulation, commonly called on off keying (OOK), is the integral of the overlap of the two functions.

$$\text{BER(OOK)} = \int_0^\infty p_I(s)\,\text{P}(s, \text{SNR})\,\text{d}s. \tag{7.54}$$

This BER can be reduced by increasing power, decreasing noise or (not very practically) trying to avoid operation in strong turbulence. Also, many other modulation techniques, such as pulse position modulation (PPM), binary phase shift keying (BPSK), quadrature phase shift keying (QPSK), single polarization state differential phase modulation (DPSK), differential quadrature phase shift keying (DQPSK) and dual polarization state differential quadrature phase shift keying (DP-QPSK) can demonstrate superior performance under some atmospheric turbulence conditions (Barua 2011, Rappaport 2009, Wang *et al* 2014). It has also been shown that scintillation-induced BER increases can be compensated with adaptive optics (Tyson 2002).

References

Abramowitz M and Stegun I A 1972 *Handbook of Mathematical Functions* (New York: Dover)

Agrawal G P 2010 *Fiber-optics Communication Systems* 4th edn (Hoboken, NJ: Wiley)

Al-Habash M A, Andrews L C and Phillips R L 2001 Mathematical model for the irradiance probability density function of a laser beam propagating through turbulent media *Opt. Eng.* **40** 1554–62

Andrews L C, Phillips R L and Hopen C Y 2001 *Laser Beam Scintillation with Applications* (Bellingham, WA: SPIE)

Anguita J A, Djordjevic I B, Neifeld M A and Vasic B V 2005 High-rate error-correction codes for the optical atmospheric channel *Proc. SPIE* **5892** 58920V

Arica N and Yarman-Vural F 2001 An overview of character recognition focused on off-line handwriting *IEEE Trans. Syst. Man Cybern.* C **31** 216–33

Arines J 2003 Least-squares modal estimation of wrapped phases: application to phase unwrapping *Appl. Opt.* **42** 3373–8

Aschenbrenner N 2010 500 megabits/second with white LED light *News Release, 18 January* Siemens Global Website

Barua B 2011 Comparison of the performance of free-space optical communication with OOK and BPSK modulation under atmospheric turbulence *Int. J. Eng. Sci. Technol.* **3** 4391–9

Bouchet O, Sizun H, Boisrobert C and de Fornel F 2006 *Free-space Optics: Propagation and Communication* (Newport Beach, CA: ISTE)

Bryngdahl O 1974 Geommetrical transformations in optics *J. Opt. Soc. Am.* **64** 1092–9

Cappellini V (ed) 1997 *Time-varying Image Processing and Moving Object Recognition* vol 4 (Waltham, MA: Elsevier)

Casasent D and Chen J 1983 Nonlinear local image preprocessing using coherent optical techniques *Appl. Opt.* **22** 808–14

Casasent D and Pasaltis D 1977 New optical transforms for pattern recognition *Proc. IEEE* **65** 77–84

Chang H T and Tsan C L Image watermarking by use of digital holography embedded in the discrete-cosine-transform domain *Appl. Opt.* **44** 6211–9

Cheriet M, Kharma N, Liu C-L and Suen C 2007 *Character Recognition Systems: A Guide for Students and Practitioners* (Hoboken, NJ: Wiley and Sons)

Collier R J, Burckhardt C B and Lin L H 1971 *Optical Holography* (New York: Academic)

Connor S 2017 Star Wars 'hologram' may become TV reality *News Item, The Independent* http://www.independent.co.uk/news/science-2124595.html

Crisp J 2005 *Introduction to Fiber Optics* 3rd edn (Oxford: Newnes)

Deans S R 1983 *The Radon Transform and Some of its Applications* (New York: Wiley)

De Caro C D, Renn A and Wild U P 1991 Hole burning, Stark effect, and data storage: holographic recording and detection of spectral holes *Appl. Opt.* **30** 2890–8

Duarte F J 2005 Secure interferometric communications in free space: enhanced sensitivity for propagation in the meter range *J. Opt. A: Pure Appl. Opt.* **7** 73–5

Duarte F J, Taylor T S, Black A M, Davenport W E and Varmette P G 2011 N-slit interferometer for secure free-space optical communications: 527 m intra interferometric path length *J. Opt.* **13** 035710

Esener S 1989 3D optical memories for high performance computing *Proc. SPIE* **1150** 113–9

Ferrari J A and Flores J L 2010 Nondirectional edge enhancement by contrast-reverted low-pass Fourier filtering *Appl. Opt.* **49** 3291–6

Ferrari J A, Flores J L and Garcia-Torales G 2010 Directional edge enhancement using a liquid-crystal display *Opt. Commun.* **283** 2803–6

Flaherty F A 2009 Physically constrained Fourier transform deconvolution method *J. Opt. Soc. Am.* A **26** 1191–4

Gabor D 1948 A new microscopic principle *Nature* **161** 777–8

Ghiglia D and Pritt M 1998 *Two-dimensional Phase Unwrapping Theory, Algorithms and Software* (New York: Wiley)

Gluckstad J and Mogensen PC 2001 Optimal phase contrast imaging in common path interferometry *Appl. Opt.* **40** 268–82

Gonzalez R C and Woods R E 1992 *Digital Image Processing* (Reading, MA: Addison-Wesley)

Goorskey D J, Schmidt J and Whiteley M R 2013 Efficacy of predictive wavefront control for compensationg aero-optical aberrations *Opt. Eng.* **52** 071418

Grubor J, Randel S, Langer K-D and Walewski J W 2008 Broadband information broadcasting using LED-based interior lighting *J. Lightwave Technol.* **26** 3883–92

Hayes J 2006 *Fiber-optics Technician's Manual* (Clifton Park, NJ: Thomson Delmar Learning)

Heanue J F, Bashaw M C and Hesselink L 1994 Volume holographic storage and retrieval of digital data *Science* **265** 749–52

Hillerkuss D *et al* 2011 26 Tbit s^{-1} line-rate super-channel transmission utilizing all-optical fast Fourier transform processing *Nature Photon* **5** 364–71

Hooke R 1665 Of a new property in the air *Micrographia* **58** 217–9

Hsu Y-N and Arsenault H H 1982 Rotation invariant digital pattern recognition using circular harmonic expansion *Appl. Opt.* **21** 4012–5

Jones R and Wykes C 1989 *Holographic and Speckle Interferometry* (Cambridge: Cambridge University Press)

Jin W M and Yan C J 2007 Optical image encryption based on multi-channel fractional Fourier transform and double random phase encoding techniques *Optik* **118** 38–41

Kishk S and Javidi B 2003 Watermarking of three-dimensional objects by digital holography *Opt. Lett.* **28** 167–9

Knight W 2005 Holographic-memory discs may put DVDs to shame *New Scientist online* November

Kong D, Shen X, Xu Q, Xin W and Guo H 2013 Multiple-image encryption scheme based on cascaded fractional Fourier transform *Appl. Opt.* **52** 2619–25

Kramer M 2013 Zap! NASA laser breaks space communications record, 23 October, www.space.com

Leith E N and Upatnieks J 1962 Wavefront reconstruction and communication theory *J. Opt. Soc. Am.* **52** 1123–30

Leith E N and Upatnieks J 1970 Wavefront reconstruction using a coherent reference beam *US Patent* No. 3506327

Li H-Y S, Qiao Y and Psaltis D 1993 Optical neural network for real-time face recognition *Appl. Opt.* **32** 5026–35

Li Y and Su X 2005 Fast algorithm for reliability-guided phase unwrapping *Opt. Electron. Eng.* **32** 76–79

Liansheng S, Meiting X and Ailing T 2013 Multiple-image encryption based on phase mask multiplexing in fractional Fourier transform domain *Opt. Lett.* **38** 1996–8

Lin C and Shen X J 2012 Analysis and design of impulse attack free generalized joint transform correlator optical encryption scheme *Opt. Laser Technol.* **44** 2032–6

Liu S T, Mi Q and Zhu B 2001 Optical image encryption with multistage and multichannel fractional Fourier domain filtering *Opt. Lett.* **26** 1242–4

Matoba O and Javidi B 2001 The keys to holographic data security: encrypted optical memory systems based on multidimensional keys for secure data storage and communication *IEEE Photon. Soc. News* October

Mazurczyk W and Szczypiorski K 2008 Steganography of VoIP Streams *Proc. 3rd Int. Symp. on Information Security (IS'08) (Monterrey, Mexico) (Lecture Notes in Computer Science* vol 5332*)* (Berlin: Springer)

Mazzaferri J and Ledesma S 2007 Rotation invariant real-time optical edge detector *Opt. Commun.* **272** 367–76

Meneses-Fabian C, Montes-Perez A and Rodriguez-Zurita G 2011 Directional edge enhancement in optical tomography of thin phase objects *Opt. Express* **19** 2608–11

Messier D 2013 NASA laser system sets record with data transmissions from Moon *Parabolic Arc* October

Nakano M and Minami T 1985 Holographic vibration analysis and a study of engine noise control *SAE Technical Paper* 85032

Ogai K, Matsui S, Kimura Y and Shimizu R 1993 An approach for nanolithography using electron holography *Japan J. Appl. Phys.* **32** 5988–92

Ou X, Horstmeyer R, Yang C and Zheng G 2013 Quantitative phase imaging via Fourier ptychographic microscopy *Opt. Lett.* **38** 4845–8

Peckham M 2013 NASA beams Mona Lisa image into space *Time* 21 January, http://newsfeed.time.com

Pepper D M 1986 Applications of optical phase conjugation *Sci. Am.* January

Peticolas F A P, Anderson R J and Kuhn M G 1999 Information hiding: a survey *Proc. IEEE* **87** 1062–78

Plamondon R and Srihari S N 2000 Online and offline handwritten character recognition: a comprehensive survey *IEEE. Trans. Pattern Anal. Mach. Intell.* **22** 63–84

Plesniak W, Pappu R and Benton S 2001 Tangible, dynamic holographic images *Three-dimensional Holographic Imaging* ed C J Kuo and M H Tsai (Hoboken, NJ: Wiley-Interscience)

Ransom P L 1980 Proposal for holographic imaging through phase-distorting media without alignment *Opt. Lett.* **5** 327–9

Rappaport T D 2009 *Wireless Communication Principles and Practice* 2nd edn (New Delhi: Pearson Education India)

Reinhardt C N, Tsintikidis D, Hammel S, Kuga Y, Ritcey J A and Ishimaru A 2012 Atmospheric channel transfer function estimation from experimental free-space optical communications data *Opt. Eng.* **51** 031205

Rice S V, Nagy G and Nartker T A 1999 *Optical Character Recognition: An Illustrated Guide to the Frontier (Springer International Series in Engineering and Computer Science)* (New York: Springer)

Saxby G 2012 *Practical Holography* 3rd edn (Boca Raton, FL: CRC Press)

Schnars U and Jüptner W 2005 *Digital Holography* (Berlin: Springer)

Schofield M A and Zhu Y M 2003 Fast phase unwrapping algorithm for interferometric applications *Opt. Lett.* **28** 1194–6

Serafino G and Sirotti P 2000 Phase image visualization with incoherent light sources *Proc. 9th Int. Symp. on Flow Visualization (Edinburgh)* pp 454–9

Serafino G and Sirotti P 2002 Phase image visualization with white light extended sources: a Fourier-optics-based interpretation *Opt. Eng.* **41** 2459–555

Serrano-Garcia D-I, Martínez-García A, Rayas-Alvarez J-A, Toto Arellano N I, Rodriguez-Zurita G and Montes Pérez A 2012 Adjustable-window grating interferometer based on a Mach–Zehnder configuration for phase profile measurements of transparent samples *Opt. Eng.* **51** 055601

Settles G S 2001 *Schlieren and Shadowgraph Techniques: Visualizing Phenomena in Transparent Media* (Berlin: Springer)

Shechtman Y, Eldar Y C, Cohen O and Segev M 2013 Efficient coherent diffractive imaging for sparsely varying objects *Opt. Express* **21** 6327–38

Shi X and Zhao D 2011 Image hiding in Fourier domain by use of joint transform correlator architecture and holographic technique *Appl. Opt.* **50** 766–72

Shi Y, Situ G and Zhang J 2006 Optical image hiding in the Fresnel domain *J. Opt. A: Pure Appl. Opt.* **8** 559–77

Shi Y, Situ G and Zhang J 2007 Multiple-image hiding in the Fresnel domain *Opt. Lett.* **32** 1914–6

Silva A, Monticone F, Castaldi G, Galdi V, Alù A and Engheta N 2014 Performing mathematical operations with metamaterials *Science* **343** 160–3

Situ G and Zhang J 2004 Double random-phase encoding in the Fresnel domain *Opt. Lett.* **29** 1584–6

Takai N and Mifune Y 2002 Digital watermarking by a holographic technique *Appl. Opt.* **41** 865–73

Teng D, Pang Z, Liu L and Wang B 2014 Displaying three-dimensional medical objects by holographical technique *Opt. Eng.* **53** 112304

Teschke M, Heyer R, Fritzsche M, Stoebenau S and Sinzinger S 2008 Application of an interferometric phase contrast to fabricate arbitrary optical elements *Appl. Opt.* **47** 2550–6

Teschke M and Sinzinger S 2009 Phase contrast imaging: a generalized perspective *J. Opt. Soc. Am.* A **26** 1015–21

Tishko T V, Dmitry T and Vladimir T 2010 *Holographic Microscopy of Phase Microscopic Objects: Theory and Practice* (Singapore: World Scientific)

Tishko T V, Titar' V P and Tishko D N 2005 Holographic methods of three-dimensional visualization of microscopic phase objects *J. Opt. Technol.* **72** 203–9

Trithemius J 1606 *Steganographia* (Frankfurt) written in 1499

Tyson R K 2002 Bit-error rate for free-space adaptive optics laser communications *J. Opt. Soc. Am.* A **19** 753–8

Tyson R K 2011a *Principles of Adaptive Optics* 3rd edn (Boca Raton FL: CRC Press)

Tyson R K 2011b Adaptive optics and ground-to-space laser communications *Appl. Opt.* **35** 3640–6

Unnikrishnan G, Joseph J and Singh K 1998 Optical encryption system that uses phase conjugation in a photorefractive crystal *Appl. Opt.* **37** 8181–6

Unnikrishnan G, Joseph J and Singh K 2000 Optical encryption by double-random phase encoding in the fractional Fourier domain *Opt. Lett.* **25** 887–9

Wang H, He J, Chen L and Tang J 2014 Blind equalization for minimum-shift keying coherent optical communication system *Opt. Eng.* **53** 046107

Wayner P 2009 *Disappearing Cryptography, Information Hiding: Steganography and Watermarking* 3rd edn (Amsterdam: Morgan Kaufmann)

Wilson C L, Watson C I and Paek E G 1997 Combined optical neural network fingerprint matching *Proc. SPIE* **3073** 373–82

Xue J *et al* 2010 An intra-chip free-space optical interconnect *37th Int. Symp. Computer Architecture*

Yaraş F, Kang H and Onural L 2010 State of the art in holographic displays: a survey *J. Disp. Tech.* **6** 443

Yariv A and Pepper D M 1977 Amplified reflection, phase conjugation, and oscillation in degenerate four-wave mixing *Opt. Lett.* **1** 16–18

Yelleswarapu C S, Kothapalli S-R and Rao D V G L N 2008 Optical Fourier techniques for medical image processing and phase contrast imaging *Opt. Commun.* **281** 1876–88

Zernike F 1942a Phase contrast, a new method for the microscopic observation of transparent objects: I. *Physica* **9** 686–98

Zernike F 1942b Phase contrast, a new method for the microscopic observation of transparent objects: II. *Physica* **9** 974–86

Zhang H, Cai L Z, Meng X F, Xu X F, Yang X L, Shen X X and Dong G Y 2013 Image watermarking based on an iterative phase retrieval algorithm and sine–cosine modulation in the discrete-cosine-transform domain *Opt. Commun.* **278** 257–63

Zheng G, Horstmeyer R and Changhuei Y 2013 Wide-field, high-resolution Fourier ptychographic microscopy *Nat. Photon.* **7** 739–45

9 780750 310567

Lightning Source UK Ltd.
Milton Keynes UK
UKOW06n1321080115

244196UK00002B/26/P

9 780750 310574